AF411199

The Infinite in Act

A Treatise on the Founding Principle of the Physical Body Endowed with Continuous Motion

Ion G. Soteropoulos

Apeiron Centre
Boston—Paris—Athens

The cover design shows the complex and indeterminate structure of the physical universe. If we perceive our universe from inside, from its center *a*, we perceive it as being an open, Euclidean plane *a* of zero curvature, which we represent by the square. If we perceive our universe from outside, from its limiting-circumference *b*, we perceive it as being a closed spherical point *b*—a singularity of infinite curvature that we represent by the circle circumscribing the square. In total, the physical universe is at any point *p* of itself both an open, infinitely extended plane *a* and a closed, extensionless point *b* without contradiction or paradox.

The logo design of the back cover is the symbol ∞, for infinity. Infinity is a symbol of impossibility. What is impossible in our finite, time-conditioned world is the continuity of our life and motion in space-time. As long as we are finite beings conditioned by time, we are programmed to interrupt our life and motion: that is, to die. We die by killing ourselves and others. It follows that because of the interruption of our life and motion, there can be no ethical act of unity, love, and justice. Thus, the statement "question the obvious, ask the impossible" is an ethical commandment to seek and realize the above impossible things, which constitute at the same time our supreme good. How it is possible to transcend our time-conditioned nature and realize such an ethical commandment is the very gnostic issue of this work.

The cover design was conceived and graphically realized by Ion Soteropoulos. The logo design was conceived by Ion Soteropoulos and graphically realized by Mathieu Pasques. They are both trademarks of the Apeiron Centre.

At age seventeen, I confronted a major existential crisis
that has determined the evolution of my life:
If life is inevitably destroyed by the sense of time,
if at the end of my life there is for me only nothingness,
then why should I continue to exist
beyond my present moment?
If there is absurd nothingness before and after life,
if life is an instant of ephemeral hope
crushed into an absurdity by two adjacent nothings,
then what is the sense of life?
This work is dedicated to those unknown heroes
who struggle to save the meaning of life
by showing that life as motion
is not an accident, a fleeting possibility, or a miracle,
but a natural and universal event
existing everywhere, at all times, under all conditions
in a perfectly uniform universe.
It follows, therefore,
that the meaning of life resides in its *continuity* in space and time.

But what is the principle of continuous life and motion?
If I continue to exist,
it is precisely because I do not function organically
according to the analytic principles of my Euclidean, particular senses
that underlie empirical science.
In fact, these analytic principles generate time,
which as a discontinuous quantity destroys any continuity of life
beyond the limit of my individual position and moment a.

What, then, is the founding principle that permits me to continue to exist
and to sustain life and motion beyond my here-and-now position at a?
What is the principle of continuous, transcendental motion
from a to something else b?

Contents

Chapter Five: The Meaning of Acceleration

Symbols and Metric Units

Symbols

< ; >	succession (before / after); asymmetric inequality; comparison (< : less than ; > : greater than); inclusion; implication (therefore: →); generation; discontinuity
=	equality; identity; simultaneity; immediacy; continuity; unity
≠	negation of equality or identity; difference; diversity
∝	proportion
×	sign of logical multiplication (product); conjunction (both); intersection; contact
+	sign of logical addition (sum); disjunction (either/or)
(′)	sign of logical negation (not)
a'	not-a: negation of a; falsification of a; the inverse or reciprocal of a ($a' = 1/a$)
b'	not-b: negation of b; falsification of b; the inverse or reciprocal of b ($b' = 1/b$)
$a = b'$	the position of a is equal to the negation of b
$b = a'$	the position of b is equal to the negation of a
1	a) finite whole 1; simple individual; sensible (organic) unit b) real, infinite whole 1; complex universe or maximum; ideal, real, physical unit c) true; necessary d) constant e) unity
∞	infinite according to extension: ∞; infinitely many; false; impossible
0	infinite according to division: $0 = 1/\infty$; nothing; false; impossible
m	mass
d	distance
t	time

F_u original force decomposed into contrary (opposite and equal) forces F and R

F force of gravity (attractive gravity); inner or centric force; curving force unifying things; force of contraction (condensation, implosion, absorption); pulling force

$R = 1/F$ force of antigravity (repulsive gravity); outer or eccentric force; stretching force separating things; force of expansion (rarefication, explosion, emission); pushing force

r radius of the circle

K curvature of the one-dimensional circle of radius r: $K = 1/r$

E energy

T temperature

P power: $P = E/t$

v speed

c speed of light

f frequency

λ wavelength

k constant of proportionality

Metric Units

cm   centimeter (length unit)

m   meter (length unit)

km   kilometer (length unit)

gm   gram (mass unit)

kg   kilogram (mass unit)

°K   degree kelvin (temperature unit)

eV   electronvolt (energy unit)

J   joule (energy unit)

W   watt (power unit)

s   second (time unit)

Y   year (time unit)

LY   light-year (length unit)

Preface

The Infinite in Act Act (τό άπειρον έν ενεργεία) is a treatise on the metaphysical foundation of the ideal, real, physical body endowed with continuous motion and on our incomplete, particular perception of the physical body.[1] The physical body may be the physical world or any member of the physical world. We define the *physical body* as the sum total of its infinite number of continuous parts. We therefore assimilate the physical body with the infinite whole. From a logical point of view, continuous motion is the continuous passage from *a* to something else *b,* despite their difference and distance. From the physical point of view, continuous motion is the power of moving everywhere, in all directions, at all times, and immediately, with a maximum speed through a finite distance in zero time or through an infinite distance in finite time.

By laying down the founding principles of the real, physical body and of our incomplete, particular perception of the physical body, we clarify the nature of their conflict. This conflict is regarded as the dominant conflict of human civilization. The treatise then proposes a solution. Indeed, assuming that the conflict and its solution have a common origin, it is sufficient to determine the nature of the conflict in order to obtain the knowledge of its solution by negation.

One manifestation of the conflict between the behavior of the real, physical body and the Euclidean, analytic structure of our particular perception is Zeno's problem of motion and the correlative problem of infinity. Infinity and continuous motion are viewed as unintelligible—that is, as absurd or impossible entities—because their synthetic and indeterminate structures violate our Euclidean, analytic perception, particularly its analytic principles of contradiction and excluded third. For example, if nothing discontinuous is continuous and if nothing

---

1. When the physical body, whose metaphysical equivalent is the physical whole, being or thing, is an object of cognition, we can call it the *physical object.*

distant is immediate, then how is it possible to have *continuous* motion between *discontinuous*, determinate parts and *immediate* motion between *distant* things? Similarly, if nothing limited is unlimited, then how is it possible to have a *finite* total sum of an *infinite* number of parts? How is it possible to have a *limited* whole whose sum of an infinite number of parts of positive magnitude is *unlimited* and that we call infinite whole? Assuming that the infinite whole and continuous, immediate motion constitute the immanent nature, the dynamic substance of the real, physical body, then their unintelligibility amounts to the unintelligibility of the physical body itself. But if it is impossible to comprehend the real, physical body through our analytic, particular experience, what is the meaning of science, of truth; and, in a general manner, what is the meaning of human life resulting from our analytic experience?

The contemporary crisis in the foundation of empirical science is the discovery that its analytic foundation is powerless to comprehend the synthetic, indeterminate nature of the real, physical body. Given the tension between the analytic structure of empirical science and the synthetic, indeterminate properties of the physical body, we have the following alternative: Either we conserve the analytic principles, particularly the analytic principle of contradiction, which Aristotle considered the most certain of all principles, at the price of destroying ontologically and epistemologically the synthetic, indeterminate, physical body or we replace the analytic principles by their negations, which are the synthetic principles taken as the immanent principles of the physical body. This treatise chooses the second alternative—getting out of the nihilistic and agnostic trap of our analytic perception in order to defend the glorious existence and intelligibility of the physical body.[2]

---

2. The ontological destruction of the physical body (claimed by ontological nihilism) consists of negating the existence of the synthetic, physical body itself, whereas the epistemological destruction of the physical body (claimed by epistemological nihilism) consists of negating the knowledge of the physical body assumed as existing but unknowable (Kantian thesis).

The gradual replacement of empirical, analytic principles by physical, synthetic principles comprises the following two stages: The first consists of showing the antinomy and conflict between the analytic principle of contradiction and the continuous motion of the physical body. Indeed, the analytic principle of contradiction judges synthetic motion as self-contradictory or paradoxical. The second consists of replacing the analytic principle of contradiction, which judges synthetic motion absurd, with its negation, which is the synthetic principle of equivalence, taken as the first founding principle of synthetic motion. We distinguish between two kinds of solutions to the problem of motion: (a) Aristotle's synthetic solution, which, insofar as it lacks the founding principle of motion, is incomplete, and (b) the complete synthetic solution postulating the founding principle of motion.

Grounded in the fundamental conflict between our analytic, particular experience and the real, physical body, we will discuss two ontological models of the body. The *synthetic* (complex) or *continuous* model displays the synthetic principles of the ideal, real, physical body, which, identified with the complex, infinite whole, is regarded as *unconditioned* and *prior* to our particular perception, and hence as being ontologically and chronologically the *first body*. We call next *synthetic, universal reason* our cognitive faculty of thinking of the body as a complex, infinite whole.

The *analytic* (simple) or *discontinuous* model displays the analytic principles of the sensible body regarded as the conditioned effect of our analytic, particular perception and hence as being posterior to our particular perception—that is, as being ontologically and chronologically a second body. We identify this incomplete, sensible body with the finite, observable portion of the physical, infinite whole, experienced by our cognitive faculties of finite, particular sensibility and analytic understanding. By perceiving the physical, infinite whole as different from what it is—namely, as being uniquely a sensible, finite part (an individual)—our particular senses suffer from perceptual illusion, and in this sense they are incomplete or imperfect. Perceptual illusion is therefore the imperceptible influence of our particular senses on the real, physical body.

Let us consider the unit distance ab ($ab = 1$), where a represents the unlimited line a_n containing an infinite multiplicity of parts and b represents the limiting-point 1 uniting all parts into one whole thing, as the magnitude of the extended, physical body and of its motion. Accordingly, if the unit distance ab vanishes, then the extended, physical body and its motion shrink to nothingness. The unit distance ab is a complex distance having the power to receive at one instant contrary determinations (senses, parts) a and b without absurdity; it also constitutes the geometric expression of the finite–infinite equivalence principle, which affirms that anything existing is both infinite and finite, an unlimited line $a = a_n$ and a limiting-point $b = 1$, infinitely many and one—that is, a complex, infinite whole.

On the other hand, let us geometrically represent the sensible body, which is the finite, observable part of the physical body, either by the unlimited line a or by the limiting-point b in conformity with the analytic, discontinuous structure $a + b$, where + stands for either/or. This latter empirical structure constitutes the *analytic principle of excluded third*, according to which we affirm that anything existing is at one instant either an unlimited line a or a limiting-point b, and is either infinite or finite. The necessity of having at one time a *unique* determination (sense) constitutes the determinate nature of the simple part. The simple part, called *individual*, which is powerless to admit at one time *contrary* determinations and hence is constrained to admit at different times *successive* determinations, constitutes the fundamental object of the Euclidean, analytic geometry.

By understanding this simple, sensible body (the individual) as different from what it appears—namely, *as if* it were the real, physical body—and the analytic principles of the simple sensible body *as if* they were the proper and true principles of the real, physical body, our cognitive faculty of understanding suffers from ontological and epistemological illusion. We note, therefore, the imperceptible influence of our particular senses on our cognitive faculty of understanding the physical body.

From the division of the body into two kinds, we derive two kinds of numbers reflecting our primitive division. The *constant* real 1 is the number of the real, physical body ruled by synthetic principles of being. The *variable* a_n is the number of the incomplete, sensible body or of the unlimited series of incomplete sensible bodies ruled by analytic principles of being and corresponding to the finite, observable portion of the physical body. Thus, the variable, sensible body a_n pictures the *analytic*—that is, the *selective, partial,* and *conflicting* manner with which we perceive the physical body—and not the physical body existing in itself (τό καθαυτό) independently of the partiality of our particular senses. We regard therefore the real 1 designating the physical body as a rest and resolution of the conflicts and absurdities generated by the infinitely regressing a_n. Indeed, only the synthetic, real 1, which is the synthetic unity of the unlimited and limited, can realize the continuous passage from the unlimited line a, representing the unlimited series a_n, to the limiting-point b, representing the real 1, and thus ground motion within the complex, physical body of magnitude $ab = 1$. Contradictions, paradoxes, or absurdities (impossibilities) do not exist in the synthetic, real 1 admitting by nature as a complex, infinite whole, spatial contrariety. They exist uniquely in our *analytic, conflicting manner* of perceiving the coexisting opposite determinations of the synthetic, real 1. Indeed, insofar as we perceive the *complex,* real 1 as different from what it *is*—a *simple,* real 1 admitting no spatial contrariety—we transform its contrary determinations, say, the unlimited line $a = a_n$ and the limiting-point $b = 1$, into conflicting opposites (contradictories) that our analytic sense constrains us to situate successively in time.

As incompletely perceiving sensible bodies (imperfect, individual observers), we live in the discontinuous, Euclidean world of our analytic, particular senses, filled with impossibilities and black holes that render continuous motion from a to something else b impossible, apparent, or incomplete. It follows that if the extended, physical world is real and that if continuous motion is a real property of the real, physical world, then the real, physical world is a continuous quantity free of impossibilities and black holes.

But what is the geometric form of the continuous, physical world? Is it unlimited or limited with respect to extent? Is it open like a flat sheet of paper or closed like a sphere? If what we experience through our analytic, particular senses is the inner, discontinuous, Euclidean part a_n of the continuous, physical whole 1, then how can we obtain access to the continuous, physical whole itself, where motion is real, complete, continuous, and immediate?

In addition to presenting a logical solution to the problem of continuous motion by identifying its founding principle, which allows the continuous passage from a_n (designating the Euclidean world of our particular senses) to 1 (designating the ideal, real, physical world independent of our particular senses), we offer geometric and empirical solutions that determine the geometric and sensuous conditions of its possibility.

The knowledge of the *first principle* that governs the real, physical body endowed with continuous motion will enable us to know the origin and solution of all our problems, our past and future, the meaning of our life and growth, and the nature of our supreme good that determines our state of maximum happiness for the maximum time as singularities, as members of our society, as members of the entire physical universe.

To facilitate the intelligibility and the conceptual accessibility of the *Inifnite in Act*, there is a glossary that defines the fundamental concepts used in this work. The definitions aim to promote the completeness and precision of the concepts under discussion.

• 1 •

The Problem of Motion

Why Motion from *a* to *b* Is Impossible

*T*he ancient Greek philosopher the Eleatic Zeno attacked the assumption that there is motion in the physical world by deriving from it absurdities.[1] Among the four arguments that Zeno posed to show that motion implying contradictions is impossible, we concentrate on the first, known as the *argument of dichotomy*. Our examination of Zeno's first argument identifies the underlying principle that he used for implicitly dismissing continuous motion as absurd or impossible (ἄτοπον). Once the principle of absurd motion is identified and then regarded as the universal source of all absurdities, it is sufficient to negate this principle in order to obtain the founding principle of continuous motion, which we understand simultaneously as the universal solution to all absurdities in the physical body. By determining the founding principle of continuous motion, we determine the founding principle of the real, physical body endowed with continuous motion. The

---

1. Motion involving the exhaustion of the inexhaustible series of halves, the immobility of the moving arrow, and the division of the indivisible instant was considered by Zeno—or by Aristotle's interpretation of Zeno—to be different manifestations of the contradictory or paradoxical nature of motion.

mystery of Nature is solved! Nature, as both a containing whole and a contained part, has the immanent property of continuous motion and is comprehensible because we know her founding principle.

In the eighth book of *Physics,* Aristotle presents two formulations of Zeno's argument of dichotomy.[2] The first formulation, which is the more familiar, states: If a body is to travel a unit distance *ab* of magnitude 1 such as *ab* = 1, it must first travel half of the distance or $1/2^1$, and then half of what remains, or $1/2^2$, and again half of what remains or $1/2^3$, and so on ad infinitum. Thus, an infinite series of halves must be traveled successively if the moving body is to reach the limit *b*. But if an infinite series is a series without limit, then how is it possible to travel an unlimited series and reach the limit *b*? It is clear that the moving body can never reach the limit *b* and hence can never move from *a* to *b* along the distance *ab* = 1. Motion is impossible because it involves a contradiction, a paradox between *ab*'s infinite series of halves deprived of limit and the imperative requirement to reach the limit *b*.

Aristotle's second formulation of Zeno's argument, which is less familiar, exposes the first argument in a different manner. Let the moving body count 1, the first half of the distance *ab* = 1, traveled in the first half of the instant; let the body count 2, the second half of what is left, traveled in the second half of the instant; let the body count 3, the third half of what is left, traveled in the third half of the instant; and so on. At the end of its travel, the moving body has counted an infinite series of whole numbers in one instant; such a case, however, is admitted to be impossible. Indeed, if nothing infinite is finite and nothing finite is infinite, then how is it possible to count an infinite series of whole numbers in a finite time—say, in one instant?[3] Similarly, if the infinite

---

2. *Physics* (VIII) 263 a 4-11 (Translation by R. Waterfield). Oxford: Oxford University Press, 1996. *Physique* (Traduction par H. Carteron). Paris: Société d'Edition " Les Belles Lettres,"1986.

3. By pairing off the half-distances and whole numbers with one another in such a way that each half-distance corresponds to a unique whole number, such

series of whole numbers determines an infinite distance, then how is it possible to draw within the finite distance $ab = 1$ an infinite distance and assign simultaneously to the same Euclidean distance ab different magnitudes—namely, finite and infinite magnitudes such as $ab = (1 = \infty)$? How is it possible that the Euclidean distance ab is an infinite whole: in other words, a limited whole that is simultaneously unlimited?

If we count the infinite number of half-distances by assigning to each half-distance a whole number, then at the end of the trip we obtain a total sum of half-distances, which is simultaneously the finite sum 1 and the infinite sum ∞. We have therefore the following table of half-distances and whole numbers in a one–to–one correspondence:

Infinite series of half-distances $\quad 0 + 1/2^1 + 1/2^2 + 1/2^3 + \ldots + 1/2^n + \ldots = 1$

Infinite series of whole numbers $\quad 0 + \quad 1 + \quad 2 + \quad 3 + \ldots + \quad n + \ldots = \infty$.

Reading horizontally, we discover the contradiction between the two unlimited series and the necessity of being limited by their respective limits and total sums. Reading vertically, the same table shows the contradictory or paradoxical case where different finite and infinite total sums are identical: $(1 \neq \infty)(1 = \infty)$.

as $1/2^n = n$ and each whole number corresponds to a unique half-distance, such as $n = 1/2^n$, we conclude that the total sum of half-distances, which is 1, and the total sum of whole numbers, which is ∞, are equivalent, such as: $1 = \infty$. By virtue of this equivalence, we can count an infinite amount of numbers in a finite time—say, in one instant. But if this equality of unequal things is a contradiction, a paradox, an absurdity, then how is it possible to count an infinite amount of numbers in one instant?

The Analytic Principle of Contradiction Declares Motion Absurd

No unlimited series of terms has a limit. According to what principle is this proposition true? By virtue of what principle is it an absurdity for the finite distance $ab = 1$ to have an infinite magnitude? The principle that forbids the finite distance to have an infinite magnitude and the unlimited series to have a limit, and thus to be traveled and counted in a finite time, is the *analytic principle of contradiction*. This principle stipulates the following: No contradictory predicates exist simultaneously in the same object perceived as a simple and indivisible substance. In other words, no contradictory predicates are conjoined or equal and hence in immediate contact.

It follows that if a real, extended, physical body—say, the unit distance ab—and its contained motion receive contradictory predicates—say, finite and infinite magnitudes—at one and the same time, then the analytic principle of contradiction qualifies the unit distance ab, its contained motion, and hence the physical body itself as self-contradictory or impossible. A fundamental epistemological conflict thus emerges between the extended, physical body receiving with respect to its unit distance *ab* and its contained motion contradictory predicates at the same time and the analytic principle of contradiction, which is incapable of comprehending the extended body—of telling us something about the nature of its unit distance *ab* and its contained motion.

Having identified the source of *ontological nihilism* (the doctrine affirming the unreality of the extended, physical body and its motion), which is the analytic principle of contradiction, we proceed to the following question:[4] What is the cognitive nature of the analytic principle of contradiction? Is its nature a priori independent of our Euclidean, particular senses or is its nature empirical, generated by our Euclidean, particular senses? Our fundamental assumption is that the analytic principle of contradiction and its correlated analytic principles of reflexive identity, excluded third, inequality, and temporal order are empirical principles reflecting the categorical and conflicting manner with which our particular senses perceive the manifold of the physical body.

Aristotle himself astutely remarked that no sense tells us that the same object admits simultaneously contradictory predicates. For example, no sense tells us that the same object *is* and *is not* at the same time.[5] This means that our particular perception is an all-or-nothing

---

4.   By *ontological nihilism* (equally called irrealism) we mean the theory asserting that nothing exists; that there is neither extended, physical body nor motion: that the extended, physical body and its contained motion are absurdities. It follows that if there is an extended, physical body and motion, then ontological nihilism necessarily predicts their violent collapse into logical contradiction and physical nothingness.

5.   Cf. *Metaphysique* Γ, 5 1010b 15-20. (Tranduction par J. Tricot). Paris: Librairie Philosophique J. Vrin, 1986. Aristotle translated the categorical and conflicting manner with which our Euclidean, particular senses experience the object into a set of analytic principles that constitute the analytic logic of the particular, sensible (observable) object displaying its analytic structure. Since that time, analytic logic is considered to be the *organon* (instrument) of *empirical science*, the science of the sensible object.

process by which we perceive either the whole object or no object at all, either the whole predicate about the object or no predicate at all.[6] From this observation, we conclude that the analytic principle of contradiction is a subjective, organic principle telling us *how* we perceive the physical object, and not *what* the physical object *really is* (τό καθαυτό) independently of our analytic, particular perception. Indeed, according to our spatial faculty of synthetic, universal reason, the real, physical object is an extended, divided, complex whole admitting at one instant, because of its extension and division, contrary determinations—that is, spatial contrariety without absurdity. Thus, relative to contrary

---

6. With our particular senses, we perceive the object as a simple, undivided whole (the individual) having at one instant a *unique* determination or *contradictory* determinations; we do not perceive the object as a complex, divided whole (the universe) having at one instant the *totality* of determinations or *contrary* determinations. To perceive the object as a simple whole means to perceive the object according to the analytic principle of excluded third—that is, either as a whole thing or as a nothing. It follows that analytic, particular perception is a categorical *all* or *nothing* process.

If at one instant we perceive the object as a simple whole a, then we assign to it the unique determination a: $a = a$, or $a = (a = a)$. This determination constitutes the analytic principle of reflexive identity, which affirms the equality of a with itself, that the whole thing a is a whole thing a. On the other hand, by virtue of the analytic principle of contradiction, it is impossible for the object perceived as a simple whole a to be equal to its negation a' (not-a): $(a = a')'$; in other words, it is impossible or false for the simple whole a to be simultaneously a whole thing a and nothing a': $a = (a = a')'$, or $a = (aa')'$. The impossible synthetic unity or conjunction of contradictory determinations forces our particular senses to select arbitrarily one of the alternative determinations. It follows that the experienced simple object is at one instant either a whole thing a or nothing a': $a + a'$, or $a = a + a'$; this constitutes the analytic principle of excluded third. Now, the simple whole a lacking its negation a' is an improper whole; in reality, it is an incomplete part. It follows that the real, proper whole is a complex thing, refuting the analytic principles of our Euclidean, analytic particular perception.

senses, the real, physical object is both a whole thing and nothing, an undivided body and an infinitely divided body, a non-extended body and an infinitely extended body, all without contradiction or paradox.

Let us take the Euclidean retina of the eye as a prototype of our incomplete, particular sense. Because the retinal cells of the eye detect only that kind of total physical light which we call *sensible* (or observable) *light* traveling in free space at the unique and finite speed c, our brain perceives only a part of the total physical object, which we call the sensible (or observable) object. The *sensible object* is a simple, undivided whole (an individual) that admits at onc instant a unique determination. We represent geometrically the simple individual either by the closed, finite point or by the open, infinite line. The logical principle expressing the sensible reality of the simple individual is the *analytic principle of contradiction*, which forbids the assignment of contrary determinations to the same simple individual at the same time. Because no contrary determinations exist simultaneously in the same individual, they are contradictory determinations existing successively in the same individual. *Temporal succession* is therefore introduced as a consequence of reducing the complex, physical whole (universe) into a simple, sensible individual incapable of admitting contrary determinations at the same time. The time-conditioned, sensible individual therefore has no self-subsistent existence apart from our analytic, particular experience. This means that if we remove the subjective constitution of our analytic, particular experience, then the relation of temporal succession between contradictory determinations of the simple, sensible individual disappears and is replaced by the complex, physical whole (universe) independent of time whose determinations are contraries existing simultaneously in space. Temporal succession, is, therefore, a contingent and apparent or illusory part of the timeless, physical whole.

The logical principle expressing the time-dependent reality of the simple, sensible individual admitting consecutive determinations at different times is the *analytic principle of inequality and temporal order*. *Particular sensibility* is our cognitive faculty for experiencing the object as a simple and time-conditioned individual receiving at one instant a unique determination and at different instants successive determinations. We call *analytic understanding* our intellectual faculty of transforming what we experience through our analytic, particular sensibility into analytic principles.

Thus, our fundamental thesis is that the origin of analytic principles is empirical, or biological. These principles reside in our incomplete, particular senses; more precisely, they reside in our low-energy and low-temperature individual brain, which perceives the complex, physical whole (universe) admitting at one instant contraries as *different* from what it is—namely, *as if* it were a simple individual admitting at one instant contradictories. The reduction of the complex, physical whole into a simple, sensible individual is called the *perceptual alteration* (corruption or degradation) of the physical whole by the process of Euclidean, analytic perception. This reduction involves the *loss* of the physical whole's comprehensive or spatial power, which comprises at one instant contrary determinations. Because we perceive the real, physical whole as different from what it is—a simple individual—we can call this perceptual alteration a *perceptual illusion*. In contrast, we call *ontological illusion* the process of taking what *appears* to our Euclidean, particular perception, the simple individual, *as if* it were the real, physical whole existing in itself (τό καθαυτό) and independently of our analytic, particular perception. A consequence of this ontological illusion is the epistemological illusion of taking the subjective, analytic principles of our particular perception *as if* they were the objective principles of the real, physical whole existing independently of our particular senses.

It is now clear that our analytic, particular experience cannot resolve the contradictions of infinity and motion and in a general manner cannot resolve the contradictions of the extended, physical body described geometrically by the unit distance *ab*, as it is this very analytic experience that produces contradictions by perceiving the physical body as a simple individual whose contradictory determinations are seen as *insolubilia.*[7]

Help, therefore, must come from a source outside and independent of our analytic, particular experience.

7. *Insolubilia* is the medieval term for insoluble contradictions. If a problem—say, a contradiction—has an impossible solution, then *insolubilia* is another way of defining the contradiction. We have contradiction when synthetic unity, or the coexistence of opposite determinations in the same thing, is impossible. The coexistence of opposite determinations is impossible and the contradiction is insoluble when we assume that the thing is a simple, undivided and unextended whole (an individual) having at one instant a unique position or contradictory positions—say, either position *a* or position *b*—and at different instants successive positions—for example, if *a*, then at a later time *b*. On the other hand, if we assume that the thing is a complex, divided, and extended whole (a universe) having at one instant contrary positions *a* and *b*, then the coexistence of opposite positions or determinations *a* and *b* in the same thing is necessary, and their contradiction is resolved into spatial contrariety.

• 2 •

Different Solutions to the Problem of Motion

Aristotle's Analytic Solution

Zeno's conclusion that the infinite extent cannot be traveled and be counted in a finite time is empirically true, because in conforming to the analytic principles of our particular experience, there is no contact or unity of the infinite with the finite, and therefore nothing infinite is finite. According to Aristotle, however, if time is infinite with respect to division, then in opposition to what Zeno concluded, it is possible for the body to travel and count the infinite extent in a finite time— say, in one instant—because the finite time itself is infinitely divided. Thus, if we divide the unit instant *ab* into an infinite number of half-instants, we obtain the analytically consistent case of a body traveling and counting an infinite series of distances and numbers ($1 + 2 + 3 + \dots$) during an infinite series of half-instants ($1/2^1 + 1/2^2 + 1/2^3 + \dots$). Accordingly, the body can travel and count the infinite extent in an infinitely divided time and not in a finite time; and if the moving body touches the infinite, it is only through the infinite and not through the finite.[1]

---

1. *Physics* (VI) 233a, 21-30

Not only is it impossible for the body to travel the infinite extent in a finite time, but it is also impossible for the body to travel the finite extent in an infinite time. But if we *infinitize* finite time and render it proportional to infinite space, or *finitize* infinite time and render it proportional to finite space, then it is possible for the body to travel the infinite extent in an infinite time, or the finite extent in a finite time: for example, in one instant.[2]

What did these two parts of Aristotle's analytic solution accomplish? What is their founding principle? The founding principle of these two Aristotelian solutions is the *analytic principle of contradiction* in association with the *analytic principles of excluded third* and *reflexive identity*. According to the analytic principle of contradiction, nothing simple admits at the same time contrary determinations. It follows that if the body and its property of motion are simple, then it is impossible for the simple body or its simple motion to receive at the same time (through its space-time factors) contrary determinations: namely, the finite and the infinite.

It follows that insofar as motion is simple, it receives at one instant, and according to the analytic principle of excluded third, either the finite or the infinite. This analytic conclusion in turn constrains us to reduce either the finite into the infinite by infinitizing all space-time factors of motion or to do the opposite. By *infinitizing* the space-time factors of motion, we assign to the body and its motion a *unique* determination—the infinite. Thus, it is analytically consistent that the body travels and counts an infinite series of distances during an infinite series of half-instants, that the body touches the infinite with the infinite, or the same with the same, in conformity with the analytic principle of reflexive identity (or reflexive contact).

Similarly, by *finitizing* the space-time factors of motion, we assign to motion a *unique* determination—the finite. It is therefore analytically consistent that the body travels and counts the finite distance $ab = 1$ in

---

2. Ibid, 237b, 23-35.

a finite time—say, in one instant—that the body touches the finite with the finite, or the same with the same, in conformity with the analytic principle of reflexive identity (or reflexive contact). Indeed, according to this reflexive principle, everything is in immediate contact with itself, and hence nothing is in immediate contact with something different from itself. This means that according to the analytic principle of contradiction, no different things (contradictories) are in immediate contact. If the thing is infinite, then the infinite thing is necessarily in immediate contact with itself, the infinite, and nothing infinite is finite. On the other hand, if the thing is finite, then the finite thing is necessarily in immediate contact with itself, the finite, and nothing finite is infinite.

Let d and t designate the space and time factors of motion; let 1 designate the finite and the necessary or consistent; let ∞ designate the infinite and the impossible; and let 0 designate nothing or impossible. If motion (speed) v is the ratio or contact of space and elapsed time:

2.1 $\qquad\qquad$ If $v = d/t$ and $d = \infty$, $t = 1$,

then a formal definition of self-contradictory or impossible motion is:

2.2 $\qquad\qquad\qquad v = \infty/1 = \infty.$

Grounded in the above formula, we assert that impossible motion consists in traveling and counting the infinite extent in a finite time. Indeed, motion that involves through its space-time factors the ratio or immediate contact of contradictories, of *infinite* distance with *finite* time, is according to the analytic principle of contradiction self-contradictory. The speed v of self-contradictory or impossible motion is infinite.

2.3 $\qquad\qquad$ If $v = d/t$ and $d = 1$, $t = \infty$,

then the reciprocal aspect of impossible motion is:

2.4 $\qquad\qquad\qquad v = 1/\infty = 0.$

Grounded in the above formula, we assert that impossible motion consists in traveling and counting the finite extent in an infinite time. Indeed, motion that involves through its space-time factors the ratio or contact of contradictories, of *finite* distance with *infinite* time, is according to the analytic principle of contradiction self-contradictory. The speed v of impossible motion is zero. Both of these formal definitions state that motion admitting at the same time (through its space-time factors) contradictory determinations—namely, the finite and the infinite—is self-contradictory or paradoxical.

Whereas the analytic principle of contradiction helps us to identify the contradictory nature of motion admitting simultaneously contradictory determinations, its correlative, the analytic principle of reflexive identity, helps us to solve motion's self-contradiction within the context of an analytic ontology of the body and its motion. Thus, if our empirical faculty of analytic understanding thinks of motion as having a simple, determinate nature, then motion is analytically consistent when we assign to it a *simple* and *unique* determination at one instant. This unique determination is either the infinite, which *infinitizes* all space-time factors of motion, or the finite, which does the opposite. Accordingly,

2.5 $$\text{if } v = d/t \quad \text{and} \quad d = \infty, \ t = \infty,$$

a formal definition of the analytically consistent and necessary motion is:

2.6 $$v = \infty \, / \, \infty = 1.$$

Grounded in the above formula, we assert that motion is analytically consistent and necessary when the body travels and counts the infinite extent in an infinite time. Indeed, motion that involves through its space-time factors the ratio or contact of the same with itself, of *infinite* distance with *infinite* time, is according to the analytic principle of reflexive identity (or reflexive contact) self-consistent. The speed of consistent and necessary motion is one.

2.7 $$\text{If } v = d/t \quad \text{and} \quad d = 1, \ t = 1,$$

then the opposite aspect of analytically consistent and necessary motion is:

2.8
$$v = 1/1 = 1.$$

Grounded in the above formula, we assert that motion is analytically consistent and necessary when the body travels and counts the finite extent in a finite time. Indeed, motion that involves through its space-time factors the ratio or contact of the same with itself, of *finite* distance with *finite* time, is according to the analytic principle of reflexive identity (or reflexive contact) self-consistent. The speed of consistent and necessary motion is one.

Because motion's internal finite/infinite contradiction destroys motion as traveling the infinite extent in a finite time or the finite extent in an infinite time, Aristotle's solution aims at liberating motion from its internal contradiction without refuting the analytic principle of contradiction, which he considered the most certain of all principles. Accordingly, on the one hand, Aristotle conserves the ontological assumption of a simple motion verifying analytic principles of organization; on the other hand, he eliminates from motion's space-time factors one of the alternative determinations. The first part of Aristotle's analytic solution eliminates the finite determination; the second eliminates the infinite determination.

Thus, what Aristotle's analytic solution effectively accomplishes is the transformation of self-contradictory motion admitting at one instant through its space-time factors contradictory determinations into an analytically consistent motion admitting through its space-time factors a unique determination, which is either the finite or the infinite. But insofar as the assignment of a *unique* determination to motion both generates and conserves the contradiction between the finite and the infinite, it follows that Aristotle's analytic solution is apparent. Indeed, the impossibility of deciding the simple nature of motion— whether simple motion has through its space-time factors a finite or an infinite determination—does not solve the finite/infinite contradiction of motion but rather conserves it. Aristotle himself recognized that although the analytic solution for resolving the problem of traveling the infinite extent in a finite time is adequate, it is not sufficient for the reality and truth of motion.[3]

But what is the reality of motion along the unit distance *ab* and the reality of the extended, physical body described by the unit distance *ab* and existing independently of our analytic, particular experience?

3. *Physics* (VIII) 263a, 11-22.

Aristotle's Incomplete Synthetic Solution

$\mathcal{Z}$eno's problem of motion along the unit distance *ab* consists of two questions. The second question, which we regard as being of secondary importance, is whether it is possible to travel and count the infinite extent in a finite time. Unlike Zeno, Aristotle answered the question positively. It is perfectly consistent to travel the infinite extent in finite time under the condition that time itself is infinitely divisible. We called Aristotle's reply to Zeno's question *the analytic solution of the problem of motion*, as his reply satisfies the analytic principles of our particular experience.

Leaving this second question aside, let's now examine the first question, which Aristotle considered more fundamental. The question is whether it is possible to traverse an infinite series of parts (both temporal and spatial) deprived of their limit and to reach their limit. For the questioner who follows Zeno's analytic reasoning, the answer is simply no! Motion is impossible because there is a contradiction between the *unlimited* series and the *limit*. Indeed, insofar as the unity or contact of the unlimited and the limited is declared by the analytic principle of contradiction to be self-contradictory, motion, as the continuous passage from the unlimited series to the limit, is equally self-contradictory.

Faced with such a powerful argument against motion and in an ultimate attempt to resolve motion's insoluble finite/infinite contradiction, Aristotle employs his ideal, spatial faculty of synthetic, universal reason—the faculty of synthetic principles. "To the question whether it is possible to traverse an infinite number of parts, either of time or of space, we must reply that in a sense it is possible and in another sense it is not possible. If the infinite exists actually (εντελεχεία), it is impossible. But if the infinite exists potentially (ἐν δυνάμει), it is possible; for in the course of continuous motion the body has traversed an infinite number of parts accidentally (κατά συμβεβηκός) but not in an absolute sense, since it is by accident that the line is an infinity of half lines, but its essence and reality are different."[4]

Even though Aristotle does not mention here to what kind of line he refers, we can surmise that he refers to the Euclidean, finite line $ab = 1$ stipulated to be actually finite with respect to extent and potentially infinite with respect to division. Clearly, if we assume (together with Aristotle) that the Euclidean, finite distance $ab = 1$ is actually, by essence and according to extension, finite and potentially by accident and according to division infinite, then it is possible for the body to travel the infinite without limit and to attain the limit. In fact, what in essence and actually the body travels is the finite with a limit, and only by accident and potentially the infinite without limit. Thus, by traveling the infinite *potentially* according to division and the finite *actually* according to extension, the mobile arrives actually and with necessity to the limit and only potentially and by accident fails to reach the limit.

We call *synthetic* this solution given by Aristotle's faculty of synthetic, universal reason. The synthetic solution aims at transforming motion's contradictory determinations, such as the finite and the infinite, into

4. Ibid, 263b, 3-8.

contraries existing simultaneously in the same motion. From the standpoint of the analytic principle of contradiction, nothing simple possesses at the same time opposite determinations; it follows that if anything simple possesses at the same time opposite determinations, then it is self-contradictory or paradoxical, and its opposite determinations are called *contradictories*. But if, as Aristotle remarks, one determination is affirmed in a certain sense (for example, potentially) and the other determination is affirmed in a different sense—that is, absolutely or actually—then it is possible that the opposite determinations exist simultaneously in the same subject without contradiction or paradox. We call these *contraries*.[5]

Thus, it is consistent to assert that the finite distance $ab = 1$ is infinite; it is consistent to assert that the body travels the unlimited series of halves and reaches their limit under the condition that we add the following: Relative to contrary senses a and a' (not-a), where a designates the actual (or essential) and a' designates the non-actual— that is, the potential (or accidental)—which we can call b, the one and the same subject—say, the distance $ab = 1$—and its contained motion— receives contrary determinations, the finite and the infinite, without contradiction or paradox. But what is the nature of the subject receiving relative to contrary senses a and $a' = b$ contrary determinations at one and the same time? It is necessarily a complex, indeterminate whole (a universe) receptive of spatial contrariety. According to Aristotle's analytic ontology of being, however, the subject's complex, indeterminate nature receiving spatial contrariety is merely of an accidental or potential nature. In fact, the subject's essential and actual nature is that of being, simple and determinate (an individual), receiving no spatial contrariety.

Can we now conclude that Aristotle has given a definitive and complete solution to Zeno's problem of motion? Certainly not! We need to recall that Aristotle solved the problem of motion under a *particular* condition. He argued that it is possible to travel the infinite series of

---

5. *Metaphysique* Γ, 6 1011b, 20-25.

halves and to reach the limit of the series under the particular condition that the traveled infinite is potential and not actual. Per contra, if the particular condition changes and the infinite series of halves is an actual infinite, then motion through the actual infinite is impossible. Indeed, how is it possible to travel *actually* an infinite series and reach *actually* the limit?

The contradiction between the infinite series without limit and the necessity of attaining the limit emerges again. This contradiction leads us to conclude that Aristotle's synthetic solution is incomplete. Motion along the unit distance *ab* is completely founded and real when it takes place under *any* condition—that is, universally—regardless of the potential or actual sense that we assign to *ab*'s infinite series of parts.

But what is the founding principle that unites the unlimited series with the limit and grounds motion as the continuous passage from the unlimited to the limit?

Analytic Convergence and the
Apparent Passage to the Limit

Zeno argued that motion from a to b is impossible because no body can travel ab's unlimited series of parts and reach the limit b. The contradiction between the unlimited series and the limit destroys any continuous passage from the unlimited to the limit and hence destroys any motion along the unit distance ab. Another factor that destroys motion from a to b is the assumption that we move algorithmically by adding successively ab's parts. We argue that no algorithmic (computational or inductive) procedure using the empirical intuition of time (succession) allows motion from a to b.

Let us try to move from a to b by adding successively ab's infinite number of halves. In the first $1/2^1$ second, we obtain the partial sum $0 + 1/2 = 1/2^1$; in the next $1/2^2$ second, we obtain the partial sum: $1/2^1 + 1/2^2 = 3/4$; in the next $1/2^3$ second, we obtain the partial sum: $3/4 + 1/2^1 = 7/8$; and so on. What we have obtained from this successive addition of the infinite number of halves is merely an infinite sequence of partial sums: $a_1 = 1/2$; $a_2 = 3/4$; $a_3 = 7/8$;... $a_n = n-1/n$;...The limit and total sum $b = 1$ is nowhere; or, to put it another way, 1 is somewhere but is inaccessible to the indefinitely approaching partial sum a_n. No matter how much we prolong the sequence of partial sums, the limit and total sum 1 will never be reached actually. In fact, the successive addition of ab's infinite number of parts produces a variable, partial sum a_n, which, insofar as it is different from the constant 1 and less than 1, such as:

2.9 $$a_n \neq 1 \quad \text{and} \quad a_n < 1,$$

it computes inexactly 1. We now comprehend clearly that no successive addition of parts produces a partial sum a_n that computes exactly— that is, with zero difference—the total sum 1; in a general manner, no algorithmic procedure using the empirical intuition of time transforms something a into something else b or allows motion from a to b.[6]

---

6. For example, let us try to move from $a = 0$ to $b = 1$ by adding successively a finite number of halves. In the first 1/2 second, we add 1/2 to 0, and we obtain the partial sum 0 + 1/2 = 1/2; in the next 1/2 second, we add 1/2 to 1/2, and we obtain the total sum 1/2 + 1/2 = 1. It seems at first that because 0 + 1/2 + 1/2 and 1 are equal, 1 is computed exactly—that is, with zero inequality—by 0 + 1/2 + 1/2. But are 0 + 1/2 + 1/2 and 1 really equal? From the Platonic point of view, they are unequal things. Indeed, 0 + 1/2 + 1/2 is a multiplicity of numbers, whereas 1 is a unique number. Thus, in reality, we must write:

$$0 + 1/2 + 1/2 < 1.$$

This means that the successive addition of a finite number of parts to the number 0 produces a finite sum 0 + 1/2 + 1/2, which is before and unequal to 1 and hence does not compute exactly 1. Now, the difference between the inequalities:

$$0 + 1/2 + 1/2 < 1 \quad \text{and} \quad 2. \quad a_n < 1$$

is the following: The first inequality is formal, because the same quantity 1 is expressed in different forms: as a multiplicity of numbers at the left side of the inequality and subsequently as a unique number at the right side of the inequality. In this sense, the problem posed by the successive addition of halves is uniquely formal—namely, how to compute exactly the unique number 1 by something different in *form*, which is the multiplicity of numbers 0 + 1/2 + 1/2. The second inequality is simultaneously *formal* and *material.* It shows that different forms a_n and 1 express different quantities—namely, the partial sum a_n varying without limit and the limited and constant total sum 1. In this

Because the positive difference $1 - a_n > 0$ indicates the impossibility of the variable, partial sum a_n to compute exactly, with zero difference, something entirely different in form and matter, which is the constant and total sum 1, we may consider the difference $1 - a_n$ as the error e of the inexact computation and description of 1 by a_n. We then call *copy, accident*, or *appearance* the variable part a_n computing and describing inexactly (incompletely) the constant, real whole 1, thought of as the *original model*, or *essence*.

The impossibility of the approximating partial sum a_n to assign the magnitude 1 to b, the impossibility of deciding whether b, the limit

sense, the problem posed by the successive addition of halves is both formal and material: how to compute exactly the total sum 1 by something different in form and matter, which is the partial sum a_n. As we note, in both cases, the successive addition of a finite or infinite number of parts gives a value a, which is different (either formally or both formally and materially) from 1. The absolute and positive difference $1 - a > 0$ renders impossible the continuous motion from a to 1, or the exact computation of 1 by a. This leads us to conclude, together with Plato, that no successive addition of parts a produces 1, the totality of parts; that we cannot generate successively a new number b from the preceding number a; that numbers are in reality *nonadditionable* (impossible to add), and timeless; that it is not time but space that is a factor of real variation within an instant; and that different numbers simultaneously occur in space since eternity.

As we discuss later in this work (see chapter 3), only if the positive difference $1 - a > 0$ is simultaneously $1 - a = 0$:

$$(1 - a > 0) = (1 - a = 0)$$

can we compute exactly 1 by something different a. But this implies that unequal things a and 1 are simultaneously equal:

$$(a < 1) = (a = 1),$$

that 1 is postulated simultaneously with a and independently of a, and in no way is 1 derived successively from a; in other words, this implies that we replace our analytic ontology of the world, where things are experienced as consecutive, simple individuals governed by analytic principles of being, with a synthetic ontology of the world, where things are contemporaneous, complex wholes (universes) governed by synthetic principles of being.

and total sum of parts, exists, whether 1 is its magnitude, and hence the impossibility of deciding whether the proposition b = the limit and total sum b exists and 1 is its magnitude is true, all indicate that the limit and total sum b = 1 is undecidable or indeterminate. Our analytic understanding now dismisses the indeterminate nature of the limit and total sum b as an absurdity, an unsolvable problem revealing either the nonexistence of b or, better, revealing a_n's ignorance of existing b, whose nature is assumed by the analytic understanding as simple and determinate receiving at one instant a unique determination.

Because it is impossible to infer successively the total sum 1 from the variable part a_n, we postulate 1 simultaneously and independently of a_n. We therefore assume that a_n and 1 are equal, and that because of their equality and simultaneity, 1 is by *direct* postulation and not by *indirect* derivation (induction or computation) the limit and total sum of the unlimited sequence of partial sums a_n. But how is it possible that two unequal things a_n < 1 are equal and simultaneous? If, because of the analytic principle of contradiction, nothing unlimited is limited, and if no whole is equal to its proper part, then how is it possible that the unlimited sequence a_n has 1 as its limit? How is it possible that the total sum 1 is equal to its partial sum a_n?

Given that the equality of unequal things violates the analytic principle of contradiction, it was imperative for the Aristotelian mathematician to find a solution to the problem of motion that does not refute the above analytic principle, considered by Aristotle to be the most certain of all principles. Such a solution was found in the *analytic theory of infinite convergent series,* which mathematicians since the time of Cauchy have developed. Insofar as this solution tried to save the absolute certainty of the analytic principle of contradiction rather to ground continuous motion, it instead constitutes an improper and apparent solution, leaving unsolvable the problem of motion, of reaching the limit 1, and hence of computing and describing exactly 1 by a_n.

The main question the analytic theory of infinite convergent series poses is how to correct the inequality $1 > a_n$ and error $e = 1 - a_n$ produced by the successive addition of the infinite number of parts without refuting the analytic principles of contradiction and temporal order. In other words, the analytic theory asks how to correct the consequences of an erroneous assumption—namely, the empirical assumption that we move algorithmically (successively) in conformity with the analytic principle of inequality and temporal order, without correcting the erroneous assumption itself.

Because the inequality and logical sequence between the terms a_n and 1 destroy inductive motion from a_n to 1 and the exact computation of 1 by a_n, it is essential to reduce the inequality and consecution into something infinitely small but not minimum. Indeed, any reduction of the inequality into zero means that the unequal terms are equal, thereby violating the analytic principle of contradiction. So when the analytic theory of convergent series reduces the inequality between the terms a_n and 1 into something infinitely small, on the one hand it conserves their inequality; on the other hand, it allows their inequality to be counted *as if* it were zero, and hence *as if* it were an equality, without committing an appreciable error. This apparent equality between the terms a_n and 1 allows us in turn to replace the variable a_n by its equal constant limit 1, and therefore to effect an apparent passage to the limit.

The Aristotelian mathematician who does not want to use the epsilon notation can use the following symbolism for expressing the above:

2.10 $$\lim_{n \to \infty} a_n = 1 \quad \text{if and only if} \quad 1 - a_n = 1/n \to 1/\infty = 0,$$

where n is a variable, finite number, which increases without limit, and ∞ is a constant, infinite number taken as the inaccessible limit of the indefinitely increasing n. This formula gives an analytic definition of the infinite convergent series. The series a_n converges (tends) to its maximum limit 1 without ever reaching 1 if, and only if, the infinitely small difference $1 - a_n = 1/n$ converges (tends) to its minimum limit $1/\infty = 0$ without ever reaching 0 as n of a_n, by passing through the infinite succession of terms, increases indefinitely without ever reaching its maximum limit ∞. Because no matter how small $1/n$ is, $1/n$ is always greater than its inaccessible minimum limit $1/\infty = 0$, we may similarly write formula 2.10 as follows:

$$2.10a \qquad \lim_{n \to \infty} a_n = 1 \quad \text{if and only if} \quad 1 - a_n = 1/n > 1/\infty = 0.$$

Now, variable quantities that are not allowed to reach their respective limits obey the axiom of Archimedes, which stipulates that there is no limit in the unlimited, Euclidean world of our analytic, particular sense.[7]

The infinitely small inequality between the consecutive terms $a_n < 1$ enables us to behave *as if* there were an equality and continuous motion between them. In reality, however, there is a persistent, discontinuous inequality that, no matter how small, destroys continuous motion. The solution of the analytic theory of convergent series is therefore apparent and incomplete; this solution attempts to correct the consequences of a false assumption but not the false assumption itself. In fact, if the source of error is our analytic, particular perception understanding the terms a_n and 1 of the unit distance ab as contradictories existing successively in time, and if no contradictories admit intermediate, continuous motion between them, then continuous motion between the consecutive terms $a_n < 1$ is impossible! It follows that the empirical assumption that we move successively, occupying different positions at different instants in conformity with the analytic principle of inequality and temporal order, is false or incomplete!

If we replace the positive infinitely small difference $1 - a_n = 1/n$ by the inequality $a_n < 1$, we can then write formula 2.10 as follows:

---

7. Indeed, according to the axiom of Archimedes, given a numeric quantity q of a given kind (no matter how small or how great q is), we can always find a quantity q' that is either less or greater than q.

2.11
$$\lim_{n \to \infty} a_n = 1 \quad \text{if and only if} \quad a_n < 1.$$

To conform to this formula, we propose an alternative analytic definition of the infinite convergent series. The series a_n converges successively to its finite limit 1 without ever reaching 1 as n of a_n increases successively to its infinite limit ∞ without ever reaching ∞ if, and only if, the terms a_n and 1 and n and ∞ verify the analytic principle of inequality and temporal order, which stipulates the inequality of unequal things. Indeed, according to this analytic principle, no matter how close to its limit the series a_n is, a_n is always before and less than its finite limit 1, such as $a_n < 1$; and no matter how great n is, n is always before and less than its infinite limit ∞, such as $n < \infty$.[8] If we regard the terms 1 and ∞ as *total* sums, which are greater than their respective *partial* sums a_n and n, then the inequalities $1 > a_n$ and $\infty > n$ verify the Euclidean whole/part inequality principle, which stipulates the inequality of the whole and its proper part. If we regard the terms 1 and ∞ as *limits,* which are greater than their respective *unlimited* series a_n and n, then the inequalities $1 > a_n$ and $\infty > n$ verify the analytic finite/infinite inequality principle, which stipulates the inequality of the limited and the unlimited.

---

8. Because the difference $1 - a_n$ is a magnitude that becomes infinitely small (a variable infinite), and the difference $\infty - n$ is a magnitude that is always equal to ∞ (a constant infinite) we conclude that ultimately the increasing a_n has an infinitely small distance from its limit 1 and the increasing number n of a_n has an infinite distance from its limit ∞. In fact, however great n is, n is as far off from ∞ as its least finite number and hence n is always a minimum part of ∞. To sum up, no matter how much the increasing a_n approaches 1, a_n is both at an infinitely small distance and at an infinite distance from 1. Clearly, not only is the indefinitely approaching a_n incapable of reaching its end and last number 1, but it also is equally incapable of moving beyond itself, which we assimilate with the least finite number a_0 of the series of numbers a_n. Continuous motion along the unit distance ab is impossible because a_n can neither begin nor end its motion!

We call *potential infinite* any of the finite variables a_n, n, and $1/n$ that indefinitely vary without reaching their respective limits, and hence without being closed, fixed, completed, or actualized by their constant limits. The potential infinite, which we assimilate with the *indefinite,* is an infinite deprived of its limit; it is a simple individual admitting at one instant, analogous to asymmetric time having a unique sense, a unique determination: the infinite. The potential infinite thereby verifies analytic principles of being, particularly the analytic principle of contradiction, which affirms that no infinite is limited, and the analytic principle of reflexive identity, which affirms that every infinite is unlimited. By having something missing, something left outside itself— namely, its limit—the indefinite or *unlimited infinite* is an incomplete, relative part with an external limit. This external limit is, by virtue of the analytic principle of inequality and temporal order, either below or beyond, either less or greater than, the indefinite part. We therefore need our empirical intuition of Euclidean time (succession) and the empirical faculty of analytic understanding to apprehend the indefinite as a time-conditioned, incomplete part geometrically represented in two dimensions by the indefinitely varying Euclidean plane.[9]

Given that continuous motion between the consecutive terms $a_n < 1$ is impossible, we conclude or predict the following: The complete, total

9. In *Physics* (cf. III, [6], 206 b, 33), Aristotle assimilates the infinite with the *incomplete part* lacking something and thus always having something external to itself. He geometrically represents this incomplete part relative to one dimension by an open, straight line to which it is always possible to add something successively. On the other hand, he assimilates the finite with the *complete whole,* which has nothing outside itself because it comprises everything and lacks nothing. He geometrically represents this complete whole with respect to one dimension by a closed circular line to which we add nothing. Cf. *Traité du Ciel* II, 4,10-25 (Traduction par J. Tricot). Paris: Librairie Philosophique J. Vrin, 1986. A two-dimensional equivalent of the incomplete straight line is the indefinitely varying Euclidean plane, and a two-dimensional equivalent of the complete circular line is the surface of a constant sphere.

sum b whose number is the real whole 1 does not exist; to put it another way, it exists, but it is unknown and inaccessible to our Euclidean, analytic, particular perception, which knows only a finite portion of the real whole 1—namely, the Euclidean, sensible part a_n. In fact, insofar as the indefinitely approaching a_n is equal to its least number a_0, our particular perception knows always, no matter how much we extend it, a minimum portion of the real whole 1. But if the ideal, real whole 1 is unknown and inaccessible to our Euclidean, analytic perception, then how do we know that 1 exists?

The answer is that we know the existence of 1 independently of our particular perception and hence rationally by our *a priori* faculties of synthetic, universal reason and transcendental imagination.[10] Thus, the ideal, real whole 1 is empirically unknown or inaccessible because it is empirically nonexistent, but it is rationally known or accessible because it is rationally existent.[11] Indeed, our ideal faculties of transcendental imagination and synthetic, universal reason postulate a priori the rational existence of 1. By means of our transcendental imagination, we intuit intellectually (formally) but not sensuously (materially) the existence of the ideal, real whole 1; and through our synthetic, universal

10. Our a priori faculty of transcendental imagination is the intuitive expression of our intellectual faculty of synthetic, universal reason. This ideal faculty has the power of intuiting intellectually (formally) the object as it *really is* (τό καθαυτό)—namely, as a timeless, physical whole (universe) numbered by the real whole 1 and independent of our particular senses. Per contra, we call particular sensibility the incomplete faculty, which intuits sensuously (materially) the object as a time-conditioned sensible part a_n (individual) generated by the analytic action of our particular sense.

11. Using Kantian terminology, we can assert that the ideal, real 1 is the intellectual object, a *noumenon*, of transcendental imagination and universal reason. The word *reason*, originating from the Latin *ratio*, corresponds to the Greek λόγος meaning "*proportionality*." As an intellectual faculty, λόγος or νούς is the capacity of thinking the synthetic unity and proportionality between different things.

reason, we translate into an idea, number, or principle what we intuit intellectually through our transcendental imagination.

Let us think through our synthetic reason the real whole 1 as a complex, indeterminate sphere of center a and radius $ab = 1$. Regarded from inside, this real whole 1 is a concave sphere, a hyperbolic surface, whose negative curvature (or repulsive force) causes the reciprocal separation between its parts a_n and 1 and hence the impossible access of 1 by a_n. Regarded from outside, this same real whole 1 is a convex sphere whose positive curvature (or attractive force) causes the reciprocal unity between its parts a_n and 1 and thus the immediate access of 1 by a_n. However, if we assume, by means of our analytic understanding, that the real whole 1 is a simple, determinate sphere having at one time a *unique* curvature, then the impossibility of determining which of the above alternative curvatures is the whole's real curvature indicates our ignorance of the simple, determinate, real whole 1. In fact, this impossibility indicates our inability to answer these analytic questions: Is the real whole 1 a sphere of positive or negative curvature? Is the real 1 a spherical or hyperbolic whole originating an attractive or repulsive force that unites or separates its parts? Is the real 1 accessible or inaccessible by its part a_n? Does the real 1 exist or not exist?

The analytic theory of convergent series was developed because of the empirical fact that we perceive the manifold of the real, physical whole 1—for instance, the multiplicity $a_n \neq 1$—discontinuously and successively according to the analytic principle of inequality and temporal order: $a_n < 1$. Because inequality and consecution between parts interrupts continuous motion, the problem of the analytic theory of convergent series is how to transform discontinuous inequality into continuous equality without violating the analytic principle of contradiction. This transformation is done by allowing the inequality $a_n < 1$ to become infinitely small so that it can be counted *as if* it were an equality. The obtained apparent equality enables us to replace a_n by its equal 1 and thus to effect the continuous passage from a_n to 1, although in reality there is neither equality nor continuous motion.

Therefore, what the analytic theory of infinite convergent series accomplishes is not the solution of the problem of motion, but merely its *regulation* by dissimulating the impossibility of solving it through analytic means. In fact, because our analytic, particular senses are incapable of perceiving the multiplicity of parts $a_n \neq 1$ simultaneously, but instead perceive them successively, generating discontinuity, inequality, and temporal order, our analytic experience, far from being the solution to the problem of motion, is its very origin. It follows that solving the problem of motion resides outside of our incomplete faculties of Euclidean, particular sensibility and analytic understanding and inside of their negations, which are the ideal faculties of synthetic, universal reason and transcendental imagination.

• 3 •

The Founding Principle of Motion

The Finite–Infinite Equivalence Principle Grounds Continuous Motion along the Unit Distance *ab*

And the ancients, superior to us and living in closer proximity to the gods, have bequeathed us this tale, that whatever is said to be consists of one and many, having in its nature limit and unlimitedness.

—Plato, *Philebus*

As we discussed in chapter 1, Zeno contended that motion along the unit distance *ab* is impossible because it generates contradictions or paradoxes such as the paradox of a body that travels an unlimited series and reaches the limit 1 and the paradox of the total sum 1 computed exactly—that is, with zero difference by something entirely different, which is the partial sum a_n. Given the unintelligibility of motion, Aristotle proposed different solutions to the paradoxes of motion. These, however, remain incomplete because they are conditional. For example, his solution of allowing the body to travel *ab*'s infinite series of parts under the condition that the series is only potentially infinite and not actually infinite is unsatisfactory because we desire to show the universality of motion—that motion is possible under any condition regardless of the nature of *ab*, whether it is finite or infinite, potentially or actually infinite.

We then turned to the modern theory of infinite convergent series. This analytic theory recognizes that continuous motion from the unlimited series a_n to its limit 1, or the exact computation of the total sum 1 by its partial sum a_n, needs an equivalence link between the different and unequal things a_n and 1 that will bridge their absolute separation and difference. Only if there is an equality between a_n and 1 can we replace a_n by its equal 1 and thus effect the continuous passage to the limit 1, or compute exactly—with zero difference—1 by a_n. But the analytic principle of contradiction judges the equality of unequal things grounding continuous motion from a_n to 1 to be absurd or impossible. Given the conflict between continuous motion and the analytic foundation of the modern theory of infinite convergent series, Aristotelian mathematicians prefer to save the analytic foundation of their theory. Accordingly, by allowing the inequality $a_n < 1$ to become infinitely small but not zero, they regard this vanishing small inequality *as if* it were an equality bridging the absolute separation of and difference between a_n and 1 without violating the analytic principle of contradiction. Because this equality dissimilates the empirical fact that there is no equality, the Aristotelian mathematicians consider it apparent. In this cunning way, the analytic foundation of the modern theory of infinite convergent series is maintained at the price of rendering motion from a_n to 1, and the equality link that grounds it, apparent and incomplete.

Both solutions, the Aristotelian and the modern, fail to solve rationally the problem of motion because they fail to identify and comprehend its founding principle, which is simultaneously the founding principle of its receptacle—the extended, physical body whose magnitude is the unit distance *ab*. This founding principle unites *ab*'s unequal and isolated parts, the unlimited series a_n and the limit 1, into a continuous, infinite whole in which continuous motion is necessary.

What our analytic understanding has dismissed as absurdity, our synthetic, universal reason considers to be the founding principle of continuous motion and its receptacle, the real, physical body. Thus, even if on the grounds of the analytic principle of contradiction there is no equality between unequal things, between the finite and the infinite, nonetheless, on the grounds of its negation, which is the synthetic principle of equivalence, we affirm the contrary. We then obtain the *finite–infinite equivalence principle*, which we take as the founding principle of continuous motion along the unit distance *ab*. The finite–infinite equivalence principle stipulates the equality between the finite and the infinite: that anything finite is simultaneously and of necessity infinite, and that anything infinite is simultaneously and of necessity finite. This equivalence principle is one of the multiple expressions of the synthetic equivalence principle, which is a principle both of the real, physical whole and of our synthetic, universal reason—the faculty of thinking of the object as a physical whole. On the grounds of the finite–infinite equivalence principle, we assert that the unlimited series a_n and the limit 1 are equal and simultaneous:

$$3.1 \qquad\qquad a_n = 1,$$

where (=) stands for unity, equality, and simultaneity.

This equality in turn enables us to replace a_n by its equal 1, and thus to effect the continuous motion from a_n to 1. But if there is an equality between a_n and 1, then a_n is simultaneously the unlimited series a_n and the limit 1 and 1 is simultaneously the limit 1 and the unlimited series a_n. This means that every member of the equivalence is a complex, infinite whole admitting at the same time opposite determinations— namely, the contraries *unlimited* and *limited*. We then write:

$$3.2 \qquad\qquad a_n = (a_n = 1) \qquad 1 = (1 = a_n).$$

At this stage we have arrived at the *Anaxagorian theory of motion*. According to the ancient Greek philosopher Anaxagoras, if there is no motion between contradictories, between nothing and something, then in order to generate something from nothing we must necessarily assume the presence of something in nothing. The same argument holds for generating nothing from something. It follows that the opposites nothing and something generate each other insofar as they are already present in each other and thus are contraries occurring simultaneously and not contradictories occurring successively.[1] Using the same Anaxagorian solution to our problem of motion, we assert the following: If there is no motion between contradictories, between the unlimited series a_n and the limit 1, then in order to generate the limit 1

---

1. For the natural philosopher Anaxagoras (fifth century BCE), mutual generation between elements is an appearance, or a delusion. In fact, for each physical body, generation is simply a process of extracting whatever already exists in the physical body considered as a universal receptacle (τό πανδεχές) or whole containing everything at all times (cf. *Physics* I (4), 187a–b). This Anaxagorian thesis leads us to conclude that generation does not produce change. Because the physical body is a synthesis of everything existing and being fixed in the physical whole since eternity, generation and destruction do not really exist in themselves independently of particular observation; in this sense, they are appearances or accidents of our particular senses (a fundamental assumption of this work). It follows, then, that *real* change must necessarily be spatial (nontemporal) assimilated to ideal locomotion, regarded as primary motion and which we define as immediate change of position within an instant, at maximum speed. The figure of ideal locomotion is the ideal rotating sphere.

By *ideal sphere* we mean the greatest of all concentric spheres of increasing finite radii comprising everything and lacking nothing, and therefore having nothing external to it. The *parts* of a rotating sphere change position; the rotating sphere taken as a *whole* occupies the same position. Thus, the rotating sphere changes position in space and at the same time occupies the same position in space (case of rest). We note that the rotating sphere is the synthetic solution to the problem of changing position without changing position, and therefore of reconciling motion with rest. The synthetic unity of motion with rest produces the rotating sphere's synthetic property of *permanent motion* of infinite duration. Thus, permanent motion is rotational and rotation (motion on a circle) is necessarily permanent and continuous.

from something entirely different, the unlimited series a_n, we must necessarily assume that 1 is already present in a_n, at least in an implicit or potential state imperceptible by our incomplete, particular senses. Aristotle would say that a_n becomes 1 if, and only if, a_n is actually a_n and potentially 1. In this sense, motion from a_n to 1 is the actualization of the potential 1 and the potentialization of the actual a_n.

The same argument holds true for the converse process: namely, for generating the unlimited series a_n from the limit 1. Now, if everything—that is, both a_n and 1—is already in everything—that is, in both a_n and 1 (as shown by the formula 3.2)—then the experienced successive motion from a_n to 1, involving a change of position at different instants, with an indefinitely increasing speed, is a perceptual illusion, an appearance, a physical impossibility. It follows that real, continuous motion independent of our analytic, particular perception must be instantaneous (immediate), involving a change of position from a_n to 1 at the same instant, with a maximum speed. In fact, this immediate motion at distance is the direct consequence of the equality and simultaneity between a_n and 1, which the finite–infinite equivalence principle stipulates.

We have argued that the analytic theory of infinite convergent series is incapable of effecting the continuous passage to the limit because it forbids, through its analytic principle of contradiction, the synthetic (complex) case of unequal things $a_n < 1$ being equal $a_n = 1$. However, what the analytic theory forbids the *synthetic* theory of infinite convergent series allows, under the condition that we replace the analytic principle of contradiction by its negation, the finite-infinite equivalence principle. Accordingly, we have *real,* continuous motion from the unlimited series a_n to the limit 1 if, and only if, on the grounds of the finite–infinite equivalence principle the unequal things $a_n < 1$ are equal and if their positive difference is at the same time equal to zero. It is therefore imperative to assign to their positive difference a zero magnitude, without destroying their positive difference, if we want to obtain real, continuous motion between *different* things. We therefore replace the analytic formula 2.10 or 2.10a (see pp. 27–28) by the following synthetic formula, which expresses the real and actual passage (μετάβασις) to the limit 1:

$$3.3 \qquad \lim_{n\,=\,\infty} a_n = 1 \quad \text{if and only if} \quad (1 - a_n > 0)\,(1 - a_n = 0),$$

where n is a variable, finite number that is equal to its limit ∞, and ∞ is a constant, infinite number taken as the accessible limit of the increasing n.

Formula 3.3 gives a synthetic definition of the infinite convergent series a_n. The series a_n converges to its maximum limit 1 and actually reaches 1 if, and only if, the positive difference $1 - a_n > 0$, which can have any value greater than zero, is simultaneously equal to zero, because n of a_n, by passing *beyond* the infinite succession of terms, attains immediately its maximum limit ∞. Because the unequal terms $a_n < 1$ are simultaneously equal $a_n = 1$, the finite limit 1 can be reached from any term a_n of the infinite series a_n. In the synthetic theory of infinite convergent series, variable quantities are at the same time constant quantities, which immediately reach their maximum limits and hence deny the empirical axiom of Archimedes that governs the Euclidean world of our analytic, particular sense.

Per contra, in the analytic theory of infinite convergent series, variable quantities obey the empirical axiom of Archimedes: They converge indefinitely to their maximum limits, and in this sense are univocally variable. Indeed, an immanent feature of the Euclidean world is the conservation of the contradiction and its derived tension between variable quantities and their inaccessible, constant limits.

Let us call *actual infinite* or *infinite in act* the real, infinite, whole 1 whose positive curvature closes, completes, fixes, and actualizes its unlimited series a_n. The infinite in act is a complex whole (universe) admitting at one instant, analogous to the circle receiving opposite senses, contrary determinations, the infinite and the finite.[2] It thereby

2. Aristotle introduced the concept of the *infinite in act* (τό ἄπειρον ἐν ἐνεργεία) in order to determine what a sensible (observable) quantity cannot be. For Aristotle, no observable quantity is actually infinite. In fact, insofar as our analytic, particular senses are actually finite, any observable quantity is actually finite and only potentially infinite. It follows that an observable quantity, which is actually infinite, is a contradiction or impossibility for our finite, particular senses. This, however, does not mean that what we cannot sense with our finite, particular senses has no physical existence independent of them. Thus, we can perfectly conceive through our synthetic, universal reason a physical quantity that is actually infinite and the same and constant for all times, and which we think of as a timeless, ideal quantity. In this case, the actual infinite is impossible with respect to our analytic, particular senses, but is necessary and consistent relative to our synthetic, universal reason, which conceives the actual infinite as an immanent property of the real, physical body, whose substance and number is the real, infinite whole 1. We must therefore distinguish between the *timeless, physical quantity*, which is the real 1 existing independently of our finite, particular senses and thought of as an infinite in act, and the *time-conditioned, sensible quantity*, which is the finite, observable part a_n of the real 1. Our finite, particular senses experience this sensible quantity as actually finite and potentially infinite.

The theory that regards the infinite whole, infinite one, or infinite in act as the real nature (substance) of the physical body is an anti-Aristotelian doctrine held by the ancient Greek philosophers (sixth to fourth centuries BCE), specifically by the Greek (Ionian) natural philosophers Anaximander, Anaxagoras, Melissus, Democritus, and Leucippus, and by the Greek philosophers Pythagoras, Zeno, and Plato. According to them, every being is both limited and unlimited, one and infinitely many, and hence a complex, infinite whole or infinite one admitting spatial contrariety. Every being is equally neither limited nor unlimited, neither one nor infinitely many, and hence is an indeterminate or impartial (neutral) *infinite one* free of the prejudice of the sensible part. The ontology that affirms the synthetic and all-comprehensive nature of the being whose substance and number is the real infinite one we call *synthetic*. From the point of view of Aristotle's analytic ontology, the above complex, verifies synthetic principles of being. For example, the infinite in act verifies the finite-infinite equivalence principle stipulating that every limited is unlimited and every unlimited is limited; it also verifies the whole-part equivalence principle, which stipulates that every whole is equal to its proper part and every part is equal to its proper whole. The assignment of a limit to the infinite that is deprived of limit transforms the infinite from an incomplete part into a complete whole, lacking nothing, called *infinite whole*. Insofar as there is nothing left outside the infinite whole, the infinite whole is an absolute and maximum quantity having no quantity either greater or smaller than itself. A maximum quantity refutes the axiom of Archimedes, the analytic principle of inequality and temporal order, and the four fundamental operations of arithmetic, which use the empirical intuition of asymmetric time for modifying successively and univocally a given quantity.

In a general manner, a thing is a maximum when it denies the analytic principles of our analytic, particular experience and when its

---

indeterminate, or impartial being is an absurd, unreal, or incomplete being because it denies analytic principles of being—say, the principle of contradiction that qualifies as absurd the unity and simultaneity of opposite determinations within the same thing. It follows that for Aristotle, the real being must obey the analytic principles of our particular experience and must be a simple and sensible (observable) being, which is either limited or unlimited, either one or infinitely many.

However, as we have argued, a simple, determinate being, which is at one time either limited or unlimited, can in no way receive continuous motion because continuous motion requires a continuity and unity between the being's parts—for example, between the unlimited series a_n and its limit 1. It follows that only a complex being, which is simultaneously unlimited and limited, has the power to receive continuous motion. If continuous motion is an immanent property of the real being, then the real being is not the simple, determinate, sensible being that is either limited or unlimited, but rather the complex, indeterminate, physical being that is both unlimited and limited, neither unlimited nor limited, and which we identify with the physical, infinite whole.

closed volume is sufficiently comprehensive to include at once the totality of its determinations—for example, its limited volume given by its positive or infinite curvature—and also its negation, the unlimited volume given by its negative or zero curvature. A closed volume having a maximum capacity of comprehension is the greatest sphere—the limiting sphere regarded as the ideal and real sphere. In that the limiting sphere contains at any time everything, the totality of its determinations, it constitutes the geometric form of the infinite whole: the infinite in act. We therefore need our ideal faculty of transcendental imagination to intuit the limiting sphere's spatial capacity of containing everything at once and also our ideal faculty of synthetic, universal reason for translating the intuited maximum comprehensive capacity of the limiting sphere into an ideal principle of maximum unity, stipulating the infinite unity of infinitely many things. As a complete whole lacking nothing and having nothing external to itself, the actual infinite is the real and proper infinite thought of as an absolute or maximum. This is the opposite of the potential infinite, which by virtue of being an incomplete part having everything left outside is a false maximum or a false infinite, which modern mathematicians call *improper infinite.*[3]

The direct consequence of postulating the finite–infinite equivalence principle as the founding principle of continuous motion is the discovery that ultimately the unlimited and the limited are different parts of one and the same continuous thing: the infinite whole, divided into infinite wholes receiving continuous and immediate motion. Far from being something unnatural (παρά φύσιν)—that is, something

---

3. The incomplete infinite deprived of its limit (that is, the Aristotelian potential infinite, which is in reality a finite magnitude varying beyond all limits) the nineteenth-century modern mathematician George Cantor called *improper infinite.* Per contra, the complete or perfect infinite (that is, the infinite, which is closed, completed, actualized, and rendered constant by the positive curvature of its spherical limit) Cantor called *proper infinite.* (See *Fondements d'une théorie générale des ensembles.* This work appeared in 1883 in the *Mathematische Annalen* XXI and in 1969 in the *Cahier pour l'Analyse: La Formalisation.*

artificial, monstrous, or ephemeral—the infinite whole, or the absolute infinite, or the infinite in act is the substance of the ideal, real, physical body numbered by the real, infinite whole 1 and figured by the limiting sphere: the sphere of spheres.

The infinite in act is not what the real, physical body abhors, but rather what our analytic, particular perception of the real, physical body repels the most.

What Magnitude Measures Exactly the Unit Distance *ab*?

So if there is a plurality, things must be both small and great; so small as to have no magnitude at all, so great as to be infinite.

—Zeno

Let us replace in formula 3.3 the positive difference $1 - a_n > 0$ by the inequality $a_n < 1$ and the zero difference $1 - a_n = 0$ by the equality $a_n = 1$. We may then write formula 3.3 as follows:

$$3.4 \qquad \lim_{n = \infty} a_n = 1 \quad \text{if and only if} \quad (a_n < 1)(a_n = 1).$$

In conformity with this formula, we propose an alternative synthetic definition of the convergent series a_n. The part a_n converges toward the whole 1 and actually reaches the whole 1 if, and only if, the part a_n, which is less than its whole 1, is simultaneously equal to its whole 1. The equality of the part and the whole we called the whole–part equivalence principle (see p. 44); this principle is an alternative expression of the finite–infinite equivalence principle. Both principles are non-Euclidean, of rational and physical origin; they ground deductive motion from the complete whole to the incomplete part, as well as the converse inductive motion from the incomplete part to the complete whole. Accordingly, the inductively converging part a_n becomes the complete whole 1, or the complete whole 1 is accessible to the inductively converging part a_n if, and only if, by virtue of the whole–part equivalence principle, the part a_n is simultaneously the whole 1 and the whole 1 is simultaneously the part a_n.

Per contra, if by virtue of the whole–part inequality principle, the whole is unequal to its proper part (as it happens in the analytic theory of infinite convergent series), we can then predict with certainty that the complete whole 1 is absolutely inaccessible to the inductively converging part a_n. We call *transcendent* 1 this absolutely inaccessible whole 1 of the analytic theory, and the relatively inaccessible and relatively accessible whole 1 of the synthetic theory of infinite convergent series we call *transcendental* 1. Insofar as the transcendent 1 is a whole, which is not itself a part, it is thus a simple whole that satisfies analytic principles of being and determines a dynamical (time-conditioned) series of parts. On the other hand, insofar as the transcendental 1 is a whole, which is simultaneously a part, it is a complex whole that verifies synthetic principles of being and determines a mathematical series of parts.[4] As a

4. Kant used the distinction between transcendent/transcendental in his *Critique of Pure Reason,* book 2: *Transcendental Dialectic,* chapter 2: The Antinomy of Pure Reason, Section IX, Solution of the Cosmological Idea of the Totality of the Dependence of Phenomenal Existences (translation by J.M.D. Meiklejohn). London: Everyman's Library, 1991. According to Kant, a *transcendent* thing is absolutely different from experience; it is an absolutely intelligible thing beyond all actual and possible experience. A *transcendental* thing, on the other hand, is both partially different from and partially identical to experience. For example, the transcendental thing may be actually intelligible and hence beyond actual experience, but it may also be potentially sensible and hence immanent to possible experience; or it may be, relative to different observers at different levels of energy and motion, both intelligible and sensible without absurdity.

Let the inductively converging a_n designate a series of sensible parts or experiences and let the constant whole 1 designate the ideal and physical total whole possessing the property of its parts, the transcendental 1 is both a containing whole and a contained part—that is, a self-contained being obeying the synthetic principle of reflexive order 1 > 1—an alternative expression of the whole–part equivalence principle.

For the total distance ab of magnitude 1 to be measured exactly—that is, with zero difference by something entirely different, the partial sum of all of its sensible parts or experiences; the 1 is a transcendent limit when it is absolutely different from and greater than its sensible part a_n, and thus when it has the Euclidean and analytic relation of inequality and temporal order with a_n, such as $a_n < 1$. Because the essence of succession is discontinuity, the relation of inequality and temporal order $a_n < 1$ destroys continuous motion between a_n and 1 and renders 1 absolutely inaccessible to a_n—that is, transcendent. The transcendent 1, which is unequal to and greater than a_n, is not a member of the series of sensible parts a_n. It follows that the ideal, transcendent 1 is a pure intelligible limit, a *noumenon,* having no sensible element. In this sense, the transcendent 1 is a simple individual, admitting at one time a unique determination—namely, the intellectual determination. Kant called *dynamical* the series of sensible parts whose intelligible limit is not a member of the sensible series (Ibid., Concluding Remark on the Solution of the Transcendental Mathematical Ideas and Introductory to the Solution of the Dynamical Ideas).

Now, the ideal, physical 1 is a transcendental limit when 1, which is greater than a_n and hence beyond experience, is equal to a_n, and hence is immanent to experience: $(1 > a_n)(1 = a_n)$. By establishing the lost equality between unequal things $a_n < 1$, continuous motion is allowed. Insofar as the transcendental 1 is beyond experience, it is an intelligible limit not belonging to the series of sensible parts (experiences) and inaccessible to the converging series a_n. Insofar as the transcendental 1 is immanent to experience, it is a sensible limit belonging to the series of sensible parts and accessible to the converging series a_n. Kant called *mathematical* the series of experiences whose intelligible limit is a member of the series of experiences and therefore is a complex limit receiving at the same time contrary determinations—namely, intellectual and sensuous determinations. In this manner, the homogeneity and continuity of the whole series is ensured and with it the continuous passage from the series of experiences to the intelligible limit of the empirical series.

distance a_n—we have assigned to their positive difference $1 - a_n > 0$ a zero magnitude that permits the unequal terms $a_n < 1$ to be simultaneously equal such as $a_n = 1$, and hence to replace a_n by its equal 1, which measures exactly with zero difference the total distance $ab = 1$.

If we count each term a_n as one, and therefore put a_n, which is the unlimited series of half-distances, into a one-to-one correspondence with n, which is the unlimited series of whole numbers, such as $a_n = n$, and if the *unlimited* series n is equal (by virtue of the finite–infinite equivalence principle) to its *limit* ∞ such as $n = \infty$, we then conclude that a_n is equal to the limit ∞: $a_n = \infty$. The unlimited series a_n therefore has two limits—the finite limit 1 and the infinite limit ∞:

3.5
$$a_n = (1 = \infty).$$

These two limits simultaneously assign two lengths or magnitudes—the finite magnitude 1 and the infinite magnitude ∞—to the same unit distance *ab*:

3.6
$$ab = (1 = \infty).$$

Because we need at the same time contrary magnitudes—the finite 1 and the infinite ∞—to measure the unit distance *ab*, we conclude that the exact measurement of the total unit distance *ab* by the partial distance a_n aims not to assign to the total unit distance *ab* a unique magnitude—say 1—but rather to realize the Platonic proportion or *synthetic unity* of opposite magnitudes, of 1 and ∞, in the same distance. It follows that the unit distance *ab* is both *one* thing—that is, a substance and the relation or principle of synthetic *unity* of different things—which we called the synthetic principle of equivalence. If the different things are opposite determinations of the same thing, such as the finite and the infinite, then the synthetic principle of equivalence becomes the specific finite–infinite equivalence principle.

In formula 3.6, let us negate (falsify) both members of the equality $1 = \infty$ in order to obtain the equivalent formula:

3.7
$$ab = (1' = \infty'),$$

where the negation (') designates *not*.

Replacing the relation of equality (=) by the equivalent relation of conjunction or contact (×), we may rewrite formula 3.7 as follows:

3.8
$$ab = (1' \times \infty'),$$

which is equal to:

3.9
$$ab = (1 + \infty)',$$

where (+) designates the disjunctive relation either/or.

Formula 3.9 gives a neutral or impartial and hence an indeterminate definition of the unit distance *ab*. It defines *ab* as that which (according to the synthetic principle of included third) has neither (not-either) the finite length 1 nor the infinite length ∞ and is neither one nor infinitely many; in other words, it defines *ab* as that which is measured neither by the finite magnitude 1 nor by the infinite magnitude ∞. In fact, the distance *ab* of magnitude 1 is a *neutral*, infinite whole, an impartial infinite one, which, by virtue of transcending the internal limitations of its parts, of the finite and the infinite, has no magnitude at all!

Because the unit distance *ab* is the finite *total* sum of an infinite series of partial sums, it is a maximum, a universe, having no magnitude beyond itself; in this sense, it constitutes the end of magnitude. As the end of magnitude, the unit distance *ab* has a zero magnitude and hence is a point—a nothing! Taking into consideration formulae 3.6 and 3.9, we conclude that the unit distance *ab* has simultaneously three lengths, or, to put it a different way, that we need at the same time three magnitudes—namely, 1, ∞, and 0—to measure exactly the unit distance *ab*. The exact definition of the unit distance *ab* is then given by the following synthetic formula:

3.10
$$ab = 1 = \infty \times 0,$$

where the unit distance *ab* equal to itself and 1 is the logical product of infinite and zero magnitudes. We can similarly define the unit distance *ab* in a Platonic manner as the unity or constant ratio (proportion, logos) of infinite and zero magnitudes:

3.11 $$ab = 1 = \infty/0.$$

Formulae 3.10 and 3.11 reveal the trinitarian nature of the unit distance *ab*, which as a maximum or uni-verse is simultaneously *one* and its *inverse*—the infinitely many and nothing.[5] The universe is therefore a *complex, infinite one* composed of contrary magnitudes, an *impartial, indeterminate one* transcending the partiality of magnitudes, and a *constant, infinite one* arising from the balance (equilibrium) of contrary magnitudes—of infinite and zero magnitudes.[6] In fact, if ∞ measures the maximum extension of the unit distance *ab* under the action of a stretching force and 0 measures the maximum division of the unit distance *ab* under the action of a contracting force, then insofar as these opposite maximum variations are equal, they are mutually neutralized, leaving the unit distance *ab* constant free of force and variation by comprising the totality of forces and variations. The unit distance

5. Uni-verse is the synthesis of the Latin words *uni*, meaning one, and *vertere*, meaning to turn. It follows that uni-verse is *one* thing, which is simultaneously its inverse *not-one*, divided into the infinitely many and nothing. In short, the proper and real universe is that which admits spatial contrariety. Because the uni-verse inverses with necessity its sense *one*, its spatial form is spherical, comprising the totality of its senses *one* and *not-one*.

6. The thesis that any extended, one thing of magnitude 1 must simultaneously have contrary magnitudes, infinite and zero magnitudes, was expounded clearly by the Greek philosopher Zeno of Elea (fifth to fourth century BCE). His argument develops as follows: Insofar as the extended, one thing of magnitude 1 is divided into an infinite number of parts, it has a zero magnitude; insofar as the total sum of an infinite number of parts of positive magnitude is infinite, the extended, one thing has an infinite magnitude. Ultimately, the extended, one thing of magnitude 1 has simultaneously infinite and zero magnitudes [see G. S. Kirk, J. E. Raven, and M. Schofield, *The Presocratic Philosophers* (Cambridge: Cambridge University Press, 1988, p. 266-269)]. How should we comprehend this affirmation? Does it show the absurd or consistent, the apparent or real nature of the extended one?

ab is therefore the synthetic magnitude, which is both the same and maximally varied according to extension and division.

If we consider ∞ to be the maximum limit of an infinite series of increasing magnitudes according to extension and 0 to be the maximum limit of an infinite series of increasing magnitudes according to division, we conclude that the unit distance *ab* has an infinite totality of lengths, that we need an infinite totality of magnitudes according to extension and division to measure exactly the unit distance *ab*.

But how should we comprehend the complex and indeterminate nature of the unit distance *ab* requiring simultaneously infinite and zero magnitudes for measuring itself as an infinite whole—an infinite one? Positively or negatively? Does the unit distance *ab* show its reality or its absurdity? Does it reveal our knowledge or our ignorance of the unit distance *ab*, of the physical body and its continuous motion?

If we assume through our empirical faculty of analytic understanding that the unit distance *ab* is a simple individual verifying analytic principles of being, then it admits at one time a unique determination; it follows, then, that the impossibility of assigning to the unit distance *ab* a unique determination (a unique magnitude: say, the magnitude 1) indicates either the absurdity and nonexistence of the unit distance *ab* or our ignorance of its real nature, which is assumed to be simple and determinate (Kantian thesis). It is clear that the impossibility of answering the simple question *What magnitude measures exactly the unit distance ab?* leads us directly to the ontological or epistemological destruction (nihilism) of the unit distance *ab*, and also of the extended, physical body and its continuous motion.[7]

---

7. In contrast to ontological nihilism (irrrealism) (see note 4, chapter 1 p. 5), the theory of epistemological nihilism (agnosticism) of the Kantian type asserts that the physical body and its continuous motion exist but that it is impossible to know their real nature, which the empirical faculty of analytic understanding assumed to be simple and determinate.

The contradiction between the analytic, ontological assumption about the simple, determinate nature of the unit distance *ab* and the impossibility of assigning to it a *unique* magnitude—say, the magnitude 1—constrains us to change the analytic, ontological assumption into its contrary. With the help of our spatial faculty of synthetic, universal reason, we assume that the unit distance *ab* of magnitude 1 is a complex, infinite whole or infinite one verifying synthetic principles of being—say, the synthetic principle of equivalence. The magnitude 1 admits therefore at any time (in conformity with the finite–infinite equivalence principle) contrary determinations—for example, infinite and zero magnitudes—without contradiction or paradox! It follows that the impossibility of deciding whether 1 measures exactly the unit distance *ab,* far from indicating the unreality or our ignorance of the unit distance *ab,* reveals its real nature and our complete knowledge, at least our intellectual knowledge, of its real nature. Indeed, assigning to the unit distance *ab* infinite and zero magnitudes, and in the overall an infinite totality of magnitudes at the same time, constitutes the exact measurement of the unit distance *ab,* which is an infinite whole and the magnitude of an infinite whole.

The change of analytic ontology and epistemology of the Aristotelian-Kantian type to synthetic ontology and epistemology of the Platonic type enables us to comprehend the unit distance *ab* and to save the extended, physical body and its motion from the destructive ontological and epistemological effects of our particular experience.

Finally, if the extended, physical body exists and if continuous motion is a real property of the extended, physical body, then their nature must be complex and indeterminate, verifying synthetic principles of being. Indeed, if the physical body were simple and determinate, then its simple nature would have contradicted its synthetic, extended nature, causing the collapse of the physical body and its motion. Thus, a physical body equal to itself and 1 has the power to exist at *a* and at the same time to continue to exist beyond *a* at *b* because since eternity the physical body is an extended body with motion whose magnitude is the synthetic, indeterminate unit distance *ab*.

The unit distance $ab = 1 = \infty \times 0$ is the exact magnitude of the ideal, real, physical body and its motion, and of any particular quantity belonging to the physical body. In other words, at their respective limits, all different, particular and variable quantities have the same, universal, common, and constant magnitude $1 = \infty \times 0$ whose number is the real, infinite whole 1.

Having determined the exact magnitude of the unit distance *ab* and thus of the extended, physical whole 1 having the unit distance *ab* as its radius or its diameter, we will proceed to derive from this exact magnitude the physical whole's exact geometry—its real and true form.

The Geometry of the Physical, Infinite Whole 1

Is the infinitely extended universe finite in size?

Let us consider the complex distance $ab = 1 = \infty \times 0$ as the radius of an ideal sphere of center a, which we take as the physical body of the real, infinite whole 1. The real, physical body of magnitude 1, defined as the limit and total sum of its infinite number of parts, may be the physical universe or any member of the physical universe.

By *ideal sphere* we mean the limiting sphere rotating at maximum speed and defined as the limit of an infinite series of Euclidean, concentric spheres of increasing finite radii and speeds. The limiting sphere is thus the greatest, highest, and fastest sphere, the sphere of spheres, the supreme sphere ruled by the synthetic principle of equivalence and zero temporal order. Because this greatest sphere contains the totality of things—namely, whatever exists and does not exist—it contains everything and thus has nothing left external to itself. In this sense, the limiting sphere is at the same time everything of infinite magnitude containing an infinity of dimensions and nothing of zero magnitude and of zero dimension, partially ending the infinite magnitude and its infinity of dimensions.

It is a transcendental, limiting sphere of maximum reality defined as the sum total of all partial realities, which we numbered by the real, infinite whole 1. Because of its complex radius $1 = \infty \times 0$, the limiting sphere brings the infinitely distant points of its infinite radius into a finite or zero distance from its center and the finitely distant points of its finite radius into an infinite distance from its center.

According to the limiting sphere's synthetic principle of equivalence, all unequal things are equal. It follows that all unequal points, directions, and dimensions of the limiting sphere are equal. This universal equality leads us directly to the *principle of perfect uniformity.* The principle of perfect uniformity states that the limiting sphere is the same everywhere at all points (homogeneity), in all directions (isotropy), in all dimensions, and at all times (eternity). The limiting sphere is therefore a constant sphere, and has a constant, positive curvature and a constant rotation at maximum speed. But what is the magnitude of the limiting sphere's positive curvature? What is the form of the limiting sphere of center a and of radius, the complex distance $ab = 1 = \infty \times 0$?

If the curvature K of a sphere is in inverse proportion to its radius r, then if the radius r of the ideal sphere is $ab = 1 = \infty \times 0$, its curvature K is $1/ab = 1 = 0 \times \infty$.[8] The inner concave part of the limiting sphere is an infinitely great sphere of infinite radius and zero curvature, whose surface is equivalent to the unlimited Euclidean or hyperbolic plane. On the other hand, the outer, convex part of the limiting sphere is an infinitely

---

8. The curvature K of a one-dimensional circle at a given point is in inverse proportion to its radius r: $K = 1/r$. The curvature K of the surface of a sphere at a given point is in inverse proportion to the square of its radius: $K = 1/r^2$. The greater the radius, the smaller the curvature. Infinite radius therefore gives a zero curvature (characterizing an open, Euclidean surface) or a negative, infinite curvature (characterizing an open, hyperbolic surface). Indeed, if zero, which is the reciprocal of the infinite, is equal to the negative infinite:
$$0 = 1/\infty = \infty',$$
then the Euclidean surface of zero curvature is equivalent to the hyperbolic surface of negative infinite curvature.

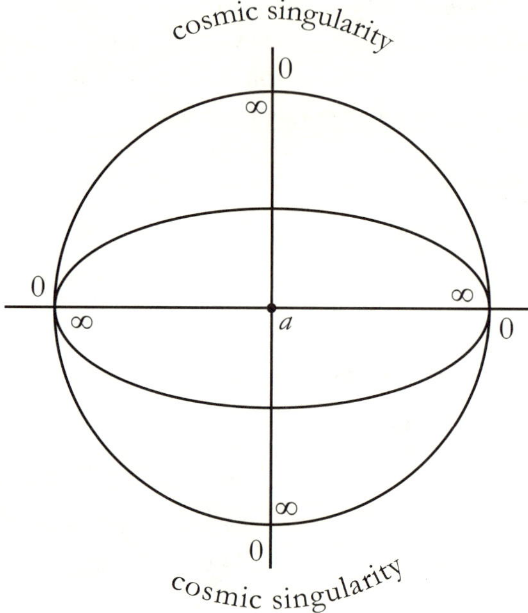

Figure 3.1. The inner concave part of the limiting sphere is an infinitely great sphere of infinite radius, whereas the outer, convex part of the limiting sphere is an infinitely small sphere of zero radius—the cosmic singularity. The concave surface of the limiting sphere curves inward, or away from the eye at the center *a*, whereas the convex surface of the limiting sphere curves outward, or toward the eye at the center *a*.

small sphere of zero radius and infinite curvature, which is equivalent to a dimensionless and extensionless point—the *cosmic singularity* (see fig. 3.1). Thus, by virtue of its complex radius and complex curvature the limiting sphere is the composition of the unlimited plane and the limiting point, the limiting point partially closing the open plane without destroying it through its infinite curvature. The limiting sphere is therefore a balanced, open-closed sphere wherein the synthetic unity and contact of opposite parts, of the unlimited and the limited, is not absurdity and chaos, but instead maximum reality and the principle of supreme being, intelligence, and order.

Indeed, as we have shown, it is the property of the limiting sphere of radius $1 = \infty \times 0$ to unify, through its zero radius, the infinitely many things of its unlimited plane without contradiction or paradox. Thus, the outer cosmic singularity is a *natural* property of the limiting sphere; it is a logical consequence of the fact that as a maximum sphere, the limiting sphere is the end of magnitude, having no magnitude beyond itself, and in this sense it is necessarily a point—a cosmic singularity! Because this outer cosmic singularity is complex and not simple, transcendental and not transcendent, it is simultaneously an inner member of the series to which it is a limiting whole without crushing the series through its infinite curvature.

A one-dimensional representation of the greatest sphere of center a and complex radius $ab = 1 = \infty \times 0$ is the greatest circle, thought of as the limiting circle or limiting circumference of an infinite series of Euclidean, concentric circles a_n of increasing finite radii. The limiting circle is formed by the complex intersection of the limiting sphere and the unlimited, Euclidean (or hyperbolic) plane passing through the limiting sphere's center a (see fig. 3.2). This intersection may be presented two-dimensionally as the composition or contact of the unlimited square with the limiting circle circumscribing the unlimited square (see fig. 3.3). The inner, unlimited Euclidean square is the two-dimensional equivalent of the inner, concave part of the limiting circle, which is the infinitely great circle of infinite radius. The horizontal axis xx and the

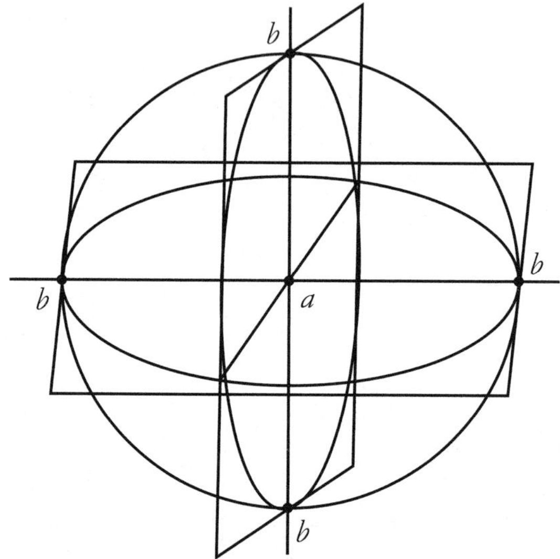

Figure 3.2. The inner concave part of the limiting sphere is an infinitely great sphere whose surface is equivalent to an unlimited plane, whereas the outer convex part of the limiting sphere is an infinitely small sphere assimilated to a spherical limiting point. On the whole, the limiting circle (or limiting circumference) is the intersection of the limiting sphere and the unlimited plane passing through the center of the limiting sphere. This means that at any point *b* of the limiting circle, the positive unit curvature $K = 1$ of the limiting circle is equal to the product of zero and infinite magnitudes: $K = 1 = 0 \times \infty$. It follows that the complex limiting circle is an infinitely curved straight line or an infinitely extended point!

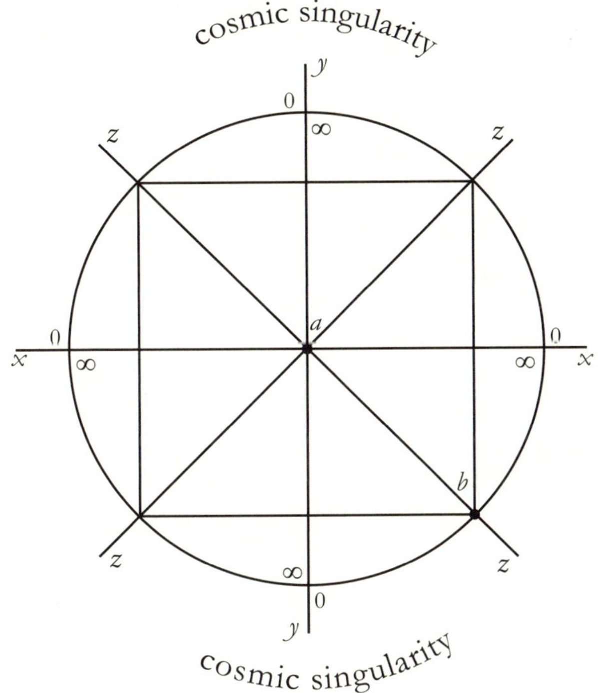

Figure 3.3. Here we see the contact of the inner, unlimited square with the outer limiting circumference circumscribing the unlimited square. The inner, unlimited square is the two-dimensional equivalent of the inner, concave part of the limiting circumference—which is an infinitely great circumference of infinite radius. The outer limiting circumference—that is, the convex part of the limiting circumference, is an infinitely small circumference of zero radius, the limiting point or cosmic singularity, circumscribing the unlimited square. The horizontal axis *xx* and the vertical axis *yy* of the limiting circumference represent, respectively, the space and time dimensions of the real, physical infinite whole; the diagonal axes *zz* represent, the proportionality between its space and time dimensions.

The real, physical, infinite whole equal to itself and 1 emerges as the synthetic unity and balance of opposites, of the unlimited square containing everything when seen from inside space-time, and of the limiting circumference containing nothing when seen from outside space-time.

vertical axis *yy* of the Euclidean square represent, respectively, the space and time dimensions of the real, physical whole 1, while the diagonal axes *zz* represent geometrically the proportionality between its space and time dimensions. On the other hand, the outer, convex part of the limiting circle is the infinitely small circle of zero radius and infinite curvature, the limiting point or *cosmic singularity* circumscribing the unlimited square geometrically representing space-time.[9]

Ultimately, the physical whole's limiting circle of radius and curvature equal to the universal and constant real $1 = \infty \times 0$ emerges as the synthetic product of contrary radii and curvatures—namely, of infinite radius and zero curvature containing everything when seen from inside and of zero radius and infinite curvature containing nothing when seen from outside. Because ∞ is the limit of an infinite sequence of magnitudes greater than 1, and 0 is the limit of an infinite sequence of magnitudes smaller than 1, we can equally resolve the complex curvature $K = 1 = \infty \times 0$ of the limiting circle into an infinite totality of curvatures K, which are both equal to 1 and greater and smaller than 1: $(K = 1)(1 < K)(K < 1)$.

Not only the physical, infinite whole 1 but also any of its parts a_n, by virtue of being equal to the real 1, is a physical, infinite whole comprising at any time the totality of radii, curvatures, geometries, and forms.

9. By the term *limiting point*, mathematicians mean any point determining an unlimited multiplicity of points. If the unlimited multiplicity of points defines an infinitely extended straight line representing one-dimensionally the infinitely extended space-time, then the limiting point is the end of the infinitely extended space-time. If the limiting point has zero extension, then it designates nothing—that is, zero space-time. On the other hand, the property of zero extension enables the limiting point to be a principle of synthetic unity unifying infinitely many and distant points of the unlimited straight line and thereby compressing everything into nothing. The curving force of this synthetic and universal unity is infinite.

Regarded as the point at which the curvature of space-time is infinite, physicists call the limiting point *space-time singularity* or *cosmic singularity*.

*In the infinitely great and small circles the center and the
circumference coincide.*

—Nicolas of Cusa

Let us now define the complex radius $ab = 1 = \infty \times 0$ of the physical, infinite whole 1 as the distance that simultaneously separates and unites the opposite points a and b, where a designates the Euclidean, inner center and lowest region and b designates the outer limiting-circumference b and highest region of the physical, infinite whole 1. Seen from inside space-time—that is, from the inner center a—the unit radius ab is a Euclidean or hyperbolic distance of infinite size, maximally separating the points a and b.

Per contra, seen from outside space-time—that is, from the outer limiting-circumference b—the same unit radius ab is a singularity of zero size maximally uniting the points a and b. It is precisely this outer cosmic singularity part of the radius ab, which, by ensuring geometrically the contact and unity of the maximally separated Euclidean points a and b, enables continuous, reversible, and immediate motion between them in the spherical body of the physical whole. Thus, if a designates the partial sum a_n and b designates the limit and total sum 1, then the continuous passage from $a = a_n$ at the Euclidean, inner center a to the outer limiting-circumference $b = 1$ is geometrically possible if, and only if, the infinite distance maximally separating a from b is simultaneously a zero distance that maximally unites a with b.[10]

This means that the continuous passage from a to b is geometrically possible if, and only if, the unit distance ab is equal to $1 = \infty \times 0$, and hence is the complex magnitude of the physical, infinite whole 1 having simultaneously infinite and zero magnitudes. It follows that Zeno's and Aristotle's conclusion, according to which it is impossible for the body

---

10. If no matter how much a_n approaches its respective limit 1, a_n is as far off from 1 as the least number a_0 of the series situated at the center a and hence equivalent to a, we can then consider a_n as equal to $a_0 = a$ and can situate it at the center a: $a_n = a_0 = a$.

to travel the spatial infinite in a finite time—say, in one instant—is false. Indeed, it is possible for the body at the center *a* to travel in one instant the infinite distance separating *a* from *b* if, and only if, the body passes *beyond* the Euclidean, infinite distance and travels through its outer cosmic singularity of zero distance connecting immediately *a* with *b*.[11]

Finally, we may think of the synthetic unit distance *ab* contacting or uniting through its cosmic singularity maximally distant points *a* and *b* as the geometric realization of the synthetic equivalence principle, which stipulates the equality of unequal points *a* and *b*: for example, of the inner center *a* and the outer limiting-circumference *b*. We obtain, then, the *center–circumference equivalence principle,* which is an alternative manifestation of the whole–part equivalence principle (see p. 44). According to this principle, the center is equal to the circumference of any concentric circle and the circumference of any concentric circle is equal to its center. It follows that the inner center *a* and the outer limiting-circumference *b* are equal and coincide.[12] Because of their equality, the contrary points *a* and *b* are everywhere at the inner center *a* and on the outer limiting-circumference *b* (see fig. 3.4):

11. Instead of using Aristotle's analytic solution of infinitizing time proportionally to infinite space by dividing the one instant into infinitely many parts in a one-to-one correspondence with spatial infinity (see pp. 13-16), we chose the geometric solution. This is a synthetic solution assigning zero magnitude, called *cosmic singularity,* to the infinite distance separating *a* from *b*. In fact, we have assumed that it is the immanent property of anything extended having magnitude and called physical body to be a complex, infinite whole equal to itself and one, and resolved into contrary infinite and zero magnitudes.

12. Nicolas of Cusa argued that in the infinitely great circle and in the infinitely small circle the center and the circumference are equal and coincide (cf Alexandre Koyré, *Du Monde Clos à L'Univers Infini* Paris: Presses Universitaires de France 1962, p. 11). Now, if the center is everywhere and the circumference is nowhere, then by virtue of the equality between center and circumference the center is globally everywhere and nowhere and the circumference is globally nowhere and everywhere.

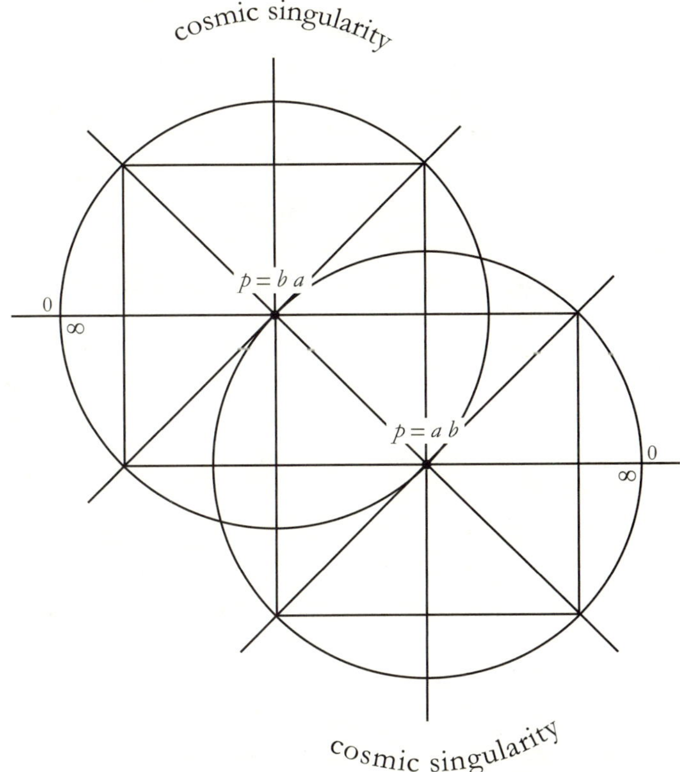

Figure 3.4. Grounded in the center–circumference equivalence principle, the inner center *a* is on the outer limiting–circumference *b* and the outer limiting–circumference *b* is at the inner center *a*. Ultimately, every point *p* of the complex body of the physical, infinite whole 1 is a complex point that is both at the center *a* inside space-time and on the limiting–circumference *b* outside space-time: $p = ab$.

3.12
$$(a = b) \rightarrow (a = ab)(b = ba).$$

Indeed, the inner center a is simultaneously at the center a and on the outer limiting-circumference b and the outer limiting-circumference b is simultaneously on the limiting-circumference b and at the center a. We conclude, then, that the physical infinite whole 1 has its inner center a and its outer limiting-circumference b everywhere, at every point p of its complex body.

If we replace in formula 3.12 ab by its equivalent $(a + b)'$ and ba by its equivalent $(b + a)'$, we obtain then the equivalent formula:

3.13
$$(a = b) \rightarrow [a = (a + b)'][b = (b + a)'],$$

stating that the contrary (equal and opposite) points a and b are nowhere—that a is neither (not-either) at the center a nor on the outer limiting-circumference b and b is neither on the outer limiting-circumference b nor at the center a.

Thus, the physical, infinite whole 1 has its inner center a and its outer limiting-circumference b nowhere in its impartial (neutral), indeterminate body. To conclude, the real, physical, infinite whole 1 has its inner center a and its outer limiting-circumference b everywhere and nowhere in its complex and indeterminate body!

Traveling, counting, and perceiving infinity within an instant by passing through the extensionless cosmic singularity encompassing the whole infinity at once constitute different manifestations of ideal or maximum motion. Physicists call this maximum motion *instantaneous action at distance*. The topos of maximum motion is the limiting-circumference b of the real, physical, infinite whole 1 occurring at a maximum distance from the fixed center a. Because motion is proportional to distance, the speed v of maximum motion is proportional to the maximum distance from the center a, which is the unit distance ab equal to the real $1 = \infty \times 0$. It follows that the speed v of maximum motion is equal to the real 1:

3.14
$$v = 1 = \infty \times 0.$$

Ideal motion of maximum speed is continuous, effortless, spontaneous, and uniform, and has a circular form; it has the same maximum speed everywhere, in all directions, at all times. Because ideal motion's maximum speed is the product of infinite and zero speeds, it is an immobile action, or an active immobility, an ενέργεια ακινησίας (to use an Aristotelian concept), where an infinite effect is produced with a finite action (or finite effort) in a finite time and a finite effect is produced with a zero action (or zero effort) in a zero time and which constitutes the principle of least action.[13] Because ideal motion occurring on the limiting-circumference b is circular, the origin and the end of ideal motion are equal and immanent to ideal motion so that both the origin and the end of ideal motion constitute ideal motion itself. Therefore, we do not need an absolutely external first or final cause, presented in the form of an Aristotelian mover, a *simple* and *transcendent mover* residing absolutely beyond the physical whole, to originate and finalize its motion, since ideal motion as circular motion is self-originated and self-finalized within the physical whole.

Finally, we call *first body* the real, physical body endowed with the power of ideal motion, which we define as the power of traveling the unit distance ab infinitely separating a from b in a finite or zero time. We can equally define the first moving body as anything having the power to produce an infinite effect with a finite or zero action in a finite or zero time and hence as anything verifying the ideal principle of least action.

By assigning a cosmic singularity of zero magnitude to the unit distance ab maximally separating any two points a and b in the Euclidean, inner part of the physical, infinite whole 1, we have universalized

---

13. According to Aristotle, an ideal being possesses its good without exercising action (see *Traité du Ciel* II 12, 292a, 20-25). This faculty of least action is the direct consequence of ideal, effortless motion where the ideal being has the power to impart movement (to itself and to something else) without being moved.

motion and rendered it immediate and possible under any condition, regardless of the magnitude of the unit distance *ab*. In a general manner, the cosmic singularity of zero magnitude ensures the permanence, uniformity, universality, and immediacy of all interactions between things independent of the magnitude of their Euclidean separation and isolation. For example, if, in conformity with Newton's inverse square law of force, gravitational attraction is inversely proportional to the square of the distance between bodies, then, if the distance is infinite, gravitational attraction is zero. In order to ensure the permanence, continuity, uniformity, and universality of gravitational attraction, the infinite distance must necessarily have a zero magnitude and therefore must be an extensionless limiting point, a cosmic singularity, when seen from outside. When distance is zero, gravitational attraction is infinite. Thus, we have the complex case wherein gravitational attraction between infinitely distant bodies is zero when seen from inside and infinite when seen from outside. On the whole, the conjunction of both magnitudes produces a finite gravitational attraction whose strength equals the real $1 = \infty \times 0$ and which is constant, uniform, universal, and immediately acting everywhere on all bodies, in all directions, at all times, and under all conditions, regardless of the distance and separation between things.

Fundamental Properties of Cosmic Singularity

We have argued that if the unit distance *ab* separating any two points *a* and *b* of the Euclidean world is not at the same time a zero distance—namely, a *cosmic singularity* when seen from outside—then it is impossible to have continuous, immediate motion between *a* and *b*. Thus, if we want to ensure the universality and permanence of continuous motion under any condition independent of the magnitude of the unit distance *ab,* this cosmic singularity must necessarily be an immanent property of the unit distance *ab* existing *simultaneously* with *ab*. It follows that a cosmic singularity, which in conformity with the temporal order of our Euclidean, analytic experience is either before or after the Euclidean unit distance *ab* but is never simultaneous with the unit distance *ab,* is powerless to ensure the unity and continuity between *a* and *b* and therefore continuous motion within the Euclidean unit distance *ab*.

A cosmic singularity preceding or succeeding the unit distance *ab* also generates the logical problem of how to derive the distance *ab* from its negation, the extensionless singularity, which is analogous to

Plato's famous logical problem of how to derive the *magnitude* from the entirely different *number,* the *line* from the entirely different *point.* Plato resolved this problem by postulating simultaneously two elements, the unextended, undivided *point* and the infinitely extended, infinitely divided *line.* From the synthetic unity of these two elements emerges Plato's *undivided line* (ἄτομος γραμμή), which enables continuous motion between the point and the line.[14] Thus, unfolding the line from

14. Plato used the complex idea of ἄτομος γραμμή or *undivided line* (point line) in order to solve geometrically the problem of deriving the magnitude from the number, the extended and divided line from the unextended and undivided point. Indeed, if the point has zero extension, then how is it possible to derive the extended line, from the unextended point? If adding points of zero extension gives a zero extension, then the only way to derive the extended line from the unextended point is to use Anaxagorian reasoning and assume that the unextended point is simultaneously an extended line and the extended line is an unextended point. The one-dimensional geometric figure arising from this assumption is the complex point line, which we can call synthetically the *extended point* or *undivided line.* The principle that permits the continuous, circular motion (circular inference) between point and line is the *point–line equivalence principle,* which is a variant of the general, synthetic equivalence principle, which governs the complex, spherical body of the physical, infinite whole 1.

Because in the Euclidean plane the point and the line are unequal and discontinuous—that is, are contradictories—any continuous motion (circular derivation) between contradictories is impossible. However, if we assume that the point and the line are equivalent in the complex surface of the sphere—that they are contraries—then continuous motion between contraries is possible. Accordingly, because the point is equivalent to the line, we replace the point by its equal line and therefore realize the continuous passage from the point to the line, from the number to the magnitude (this is the solution to Plato's problem). Conversely, because the line is equivalent to the point, we replace the line by its equal point and therefore effect the continuous passage from the line to the point, from the unlimited line a_n to the limiting-point 1 (this is the solution to the problem of effecting the continuous passage to 1; see p. 39).

the point (cosmic singularity) under the expanding action of repulsive gravity and then folding or curving the line into the point (cosmic singularity) under the contracting action of attractive gravity is possible if, and only if, at any time the unit distance $ab = 1 = \infty \times 0$ is the constant magnitude of the physical, infinite whole 1. This means that the constantly rotating sphere of the physical, infinite whole 1 appears to our particular sensibility and analytic understanding *as if* it were a Euclidean, unlimited sequence a_n of expansions and contractions, although in reality it is a timeless, constantly rotating spherical whole independent of the infinite sequence of expansions and contractions by comprising at all times the totality of mutually neutralized expansions and contractions.

In fact, the constantly rotating sphere of the physical, infinite whole 1 of constant magnitude equal to $1 = \infty \times 0$ arises as an impartial equilibrium state between opposing forces: between the stretching force of repulsive gravity expanding the physical, infinite whole 1 and the unit wavelength λ ($\lambda = 1$) of its light into a line of infinite magnitude, and the compressing force of attractive gravity contracting the physical, infinite whole 1 and the unit wavelength λ of its light into a point of zero magnitude. This balance between the contrary forces and magnitudes produces on the whole a permanently rotating physical, infinite whole 1 free of variation and force by comprising the totality of variations and forces.

Five fundamental properties characterize the cosmic singularity occurring simultaneously with the physical, infinite whole 1 and constituting the outer, infinitely convex part of its spherical body. These properties are unity, limitation, constancy, compression, and substance.

The geometric and material condition of the formal principle of equivalence is the cosmic singularity itself, which, by virtue of its zero size and infinite curvature, unites Plato's elements, the point and the line, independently of their maximum difference.

Unity

The fundamental property of unity poses the logical problem of how to eliminate the distance *ab* separating any two parts *a* and *b* of the physical, infinite whole 1 without suppressing the distance itself and hence without suppressing the body of the physical whole whose magnitude is the distance *ab*. For example, the realization of unity between *a* and *b* requires the end of the distance separating *a* from *b*. If anything having distance is an extended body, then the end of the distance is tantamount to the end of the extended body. We then have the logical problem of how to eliminate the distance between the parts of the body and thus realize their unity and immediate contact without eliminating the extended body itself.

The solution is given geometrically by the limiting sphere having as magnitude (radius or diameter) the complex unit distance $ab = 1 = \infty \times 0$; this *unextended distance* is, from the inside, an infinitely extended distance containing the spherical body's infinite multiplicity of parts and, from the outside, an unextended point without parts, a cosmic singularity, uniting through its property of unextended unity the spherical body's infinite multiplicity of parts. Thus, in order for the cosmic singularity to assign unity and contact to any two different, distant parts *a* and *b* of the spherical body that will allow continuous, immediate communication between them, it is imperative that this cosmic singularity occurs *simultaneously* with the unit distance *ab*. Only a *complex* and *transcendental* cosmic singularity that is both different from or external to the unit distance *ab* and identical or immanent to the unit distance *ab* is constructive! If a cosmic singularity is *simple* and *transcendent*—that is, *entirely* different from or external to the unit distance *ab*, and hence occurring before or after the unit distance *ab*—it is destructive, involving the annihilation of the unit distance *ab* and thus of the extended, physical body.

By assigning an unextended unity to the extended unit distance *ab*, transcendental cosmic singularity enables us to consider the physical body of magnitude $ab = 1 = \infty \times 0$ as an unextended 1 containing an infinity of extended parts according to extension and division without contradiction or paradox. The principle stipulating the unity of the physical body's infinity of parts, that *all is one* regardless of the distance within the parts and between the parts, is the *synthetic equivalence principle.* The synthetic equivalence principle is an intellectual principle of our synthetic, universal reason, which the cosmic singularity actualizes within the physical body. It follows that the synthetic equivalence principle ceases to be a subjective, regulative or intellectual principle of synthetic, universal reason and becomes an objective, constitutive or physical principle of the physical body—the physical whole 1. Instead of intellectually uniting our particular cognitions of the physical body (Kantian thesis), this synthetic equivalence principle unites directly and effectively, by means of the body's cosmic singularity, its infinite number of parts. Because synthetic, universal reason, the faculty of synthetic ideas and principles, is simultaneously, through its equivalence principle, an immanent, constitutive element of the physical body, we obtain the synthetic rationality of the physical body and the objective, physical reality (corporeality) of synthetic reason.

When different parts of the physical, infinite whole 1 are united and put into immediate contact by the cosmic singularity, they acquire common and universal features that render them uniform regardless of their difference and distance. For example, the equivalence $a = b$ permits to interchange the particular properties a and b of the left and right parts a and b so that they become common and universal properties that globally render the left and right parts uniform regardless of their particular difference: $a = b \rightarrow ab = ba$. In this manner, the *same* universal principles can determine *different* parts without absurdity. The reason is that these parts are globally equivalent uniform wholes (universes) united by a common point, the cosmic singularity. The necessary condition of the universality of the principles is, then, the unity and uniformity of the physical whole realized by its cosmic singularity.

If we want to establish completely the concreteness and actuality of the synthetic equivalence principle, it is not sufficient to show its geometric and physical existence as through the mechanism of the unifying cosmic singularity. We must also show the empirical existence of the synthetic equivalence principle as an object of sense; we must show that it exists as an object of infinite, universal sense, and that we possess the faculty of infinite, synthetic, universal sensibility for experiencing it. Only if we succeed in perceiving the cosmic singularity as effectively unifying the physical, infinite whole will we have experienced completely and maximally the unity and continuity of its body, the uniformity and universality of its principles and laws. Only then will we have experienced its principle of universal equivalence and zero temporal order, which stipulates the equality of unequal things, of everything with everything at all times, which is itself the principle of supreme beauty, intelligence, and truth, of supreme justice, love, and good, as well as the principle of continuous life and motion.

The next chapter shows under what conditions our finite brain may possess the faculty of infinite synthetic, universal sensibility—the faculty enabling us to experience the cosmic singularity of the physical whole that grounds the unity, continuity, and uniformity of its body as well as the continuous and immediate communication of its parts. Before arriving at the subsequent chapter, however, we will discuss the remaining four properties of transcendental cosmic singularity: limitation, constancy, compression, and substance.

Limitation

By virtue of being the end of an infinite sequence a_n of *discontinuous* parts, the infinitely curved cosmic singularity closes and delimits this infinite sequence in order to transform it into an infinite whole. This infinite whole is defined as the limit and total sum 1 of an infinite multiplicity of *continuous* parts. Assigning an accessible cosmic singularity or limiting point—say, 1—to the infinite, Euclidean series of parts a_n has the effect of completing and actualizing the whole's infinity, which otherwise deprived of an accessible limiting point would have remained incomplete, improper, or potential.[15] In addition to completing the incomplete infinite—that is, the indefinite a_n—the limiting–point 1 assigns unity, continuity, and simultaneity to a_n's discontinuous, consecutive parts, allowing them to communicate immediately, independent of their discontinuity and consecution.

Given the relative and not absolute character of the limitation and closure of the unlimited series of parts a_n, the physical infinite whole 1 continues to exist beyond and after its limitation 1, in conformity with the synthetic principle of reflexive order: 1 < 1.

Constancy

If motion is an immanent property of distance, then the end of distance implies the end of motion—that is, constancy or rest. It follows that the cosmic singularity, which is the end of distance and hence the end of motion, assigns constancy to the indefinitely accelerating Euclidean plane a_n and its indefinitely varying magnitude, speed, and laws. In other words, at the outer convex part of the limiting-circumference b, which is the cosmic singularity, the indefinitely accelerating Euclidean plane a_n is transformed into a constantly rotating limiting-sphere 1 of constant magnitude, speed, and laws whose flow of time is immobilized by its outer cosmic singularity.

Thus, the end of Euclidean, particular acceleration at the outer limiting-circumference b determines the emergence of an ideal, circular motion of maximum speed equal to the real 1, which we defined in formula 3.14 (see p. 66) as the product of infinite and zero speeds. Because circular motion at the maximum speed $1 = \infty \times 0$ is the end of rectilinear, particular acceleration, it is an absolute, uniform, universal,

---

15. As we have mentioned in Analytic Convergence and the Apparent Passage to the Limit, an infinite series of parts a_n deprived of its limit 1 or with an absolutely inaccessible and transcendent limit 1 is a variable magnitude that increases indefinitely but remains finite and equal to the least magnitude of the unlimited series. We conclude, then, that the infinite without limit or with an absolutely inaccessible limit is an improper infinite (Cantor), or that it is a potential infinite (Aristotle), which in reality or actually is finite.

and timeless motion transcending all particular accelerations (expansions or contractions) experienced in a conflicting or successive manner.

We may define *circular motion* (rotation or vibration), which constitutes the end of rectilinear particular acceleration, as the balance of contrary (opposite and equal) forces causing contrary accelerations. A balance of contrary, mutually neutralized forces does not result in a simple body at rest, as Aristotle thought; rather, this balance results in a complex body having simultaneously rest and constant motion and in a neutral, indeterminate body having neither rest nor constant motion. Thus, circular motion (rotation or vibration) is the synthetic, impartial solution to the problem of how to move constantly while occupying simultaneously the same position, and therefore being at rest, without contradiction or paradox.

Indeed, any extended body b on the limiting-circle b and in circular motion about a fixed center a is a complex, limiting sphere whose complex unit mass m is equal to itself and 1 divided into contrary (opposite and equal) gravitational and inertial masses.

The gravitational mass m_g originates the inward (centric) and vertical force of attractive gravity F constraining the body to move inward and its limiting-circumference b to curve toward the center a and be convex. Anything accelerating and curving toward the inner center a is low and heavy. The inward (centric) acceleration reducing the distance from the inner center a we call, depending on the case, *contraction* or *regression*. Because attractive gravity shrinks the distance from the center a—that is, decreases the radius ab of the limiting-sphere b of center a—we may regard attractive gravity as a contracting or curving force that produces concentration, indivision, unity, and contact within the limiting sphere.

The inverse inertial mass m_i, regarded as resistance to gravitational mass, originates the outward (eccentric) and horizontal force of antigravity or repulsive gravity R constraining the body to move outward and its limiting–circumference b to curve away from the center a and be concave. Anything accelerating and curving away from the inner center a is high and light. The outward (eccentric) acceleration generating the distance from the low center a we call, depending on the case, *expansion* or *progression*. If the distance from the center a is the radius ab of the limiting-sphere b of center a, then we may consider repulsive gravity as a stretching or negatively curving force producing extension, division, multiplicity, and separation within the limiting sphere. Because of the equality between the center a and the limiting-circumference b, the mass of any extended body at the center a has globally the properties of the limiting-sphere b on the limiting-circumference b and originates a complex gravitational force F_u divided into equal and opposite forces F and R whose balanced and constant ratio is equal to 1 (see fig. 3.5):

$$3.15 \qquad F_u = (F = R) \quad \text{or} \quad F_u = F / R = 1.$$

We call the above complex gravitational force, which is proportional to the complex spherical mass (or to the complex limiting-circumference b of the complex spherical mass), *original* or *first force*. We may then say that the first gravitational force F_u acts both ways (symmetrically)— namely, inward, toward the center a as attractive gravity F, and outward, away from the center a as repulsive gravity R. If we negate both contrary forces F and R, we may also say that the same gravitational force F_u acts neither (not-either) as attractive gravity F nor as repulsive gravity R,

$$3.16 \qquad F_u = (F' = R') = (F + R)',$$

and that F_u is an indeterminate and impartial gravitational force.

Ultimately, the limiting-sphere b whose complex mass (or complex limiting-circumference b of the complex mass) originates the complex gravitational force $F_u = (F = R)$ has at any time the power to do anything or everything, which we call *omnipuissance*. For example, the limiting-sphere b has at any time the power to accelerate both toward and away from the center a and thus to move permanently in a circular line at a constant distance from the center a. The same limiting sphere likewise has at any time the power to accelerate neither toward nor away from the center a and thus to be at rest at a constant distance from the center a. Omnipuissance is therefore an immanent property of the limiting-sphere b in permanent rotation and rest on the limiting-circle b.

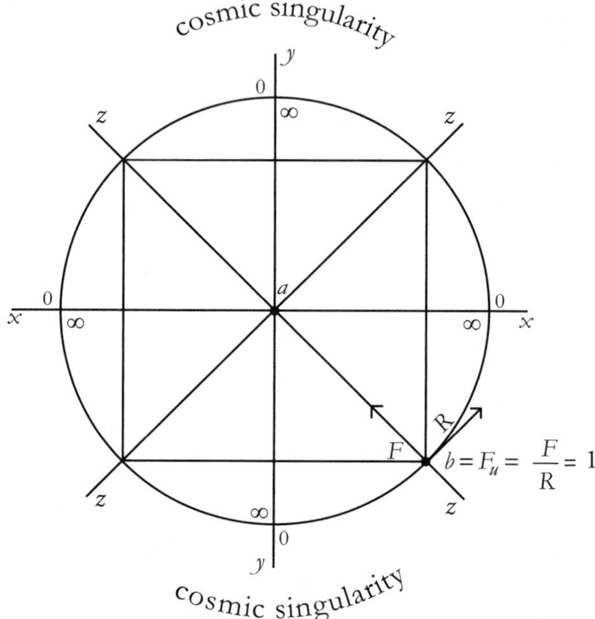

Figure 3.5 Any extended body *b* on the limiting-circle *b* is a limiting-sphere *b* whose complex mass (or complex limiting-circumference *b*) originates a complex, gravitational force F_u resolved into contrary forces *F* and *R* whose constant ratio is equal to 1. The inward vector *F* directed toward the center *a* is the curving force of attractive gravity. The outward vector *R* directed away from center *a* is the stretching force of repulsive gravity. Mutually neutralized they produce a force-free body in constant rotation. The limiting-circle *b* is both the path of the limiting sphere and the extremity of its body—its limiting-circumference *b*.

Let us define the original gravitational force $F_u = (F = R)$ as the sum and product of its contrary (equal and opposite) forces F and R:

$$3.17 \qquad F_u = (F + R = 0)(F \times R = 1).$$

This formula shows that the original force F_u emerging from the equality of opposite forces is simultaneously a zero force, when we consider it the algebraic sum of contrary forces, and a constant force equal to 1, when we consider it the product of contrary forces. This means that the state of zero force implies not the absolute absence of forces, but instead their relative and partial absence. Indeed, we have the absence of forces if, and only if, we have the constant presence of contrary forces within the complex, original gravitational force F_u. It follows that circular motion at the maximum speed $1 = \infty \times 0$ is constant and free motion independent of the constraint of an external force by virtue of being the *effect* of contrary forces or of an immanent, original force F_u resolved into contrary forces F and R. A free and constantly rotating body is the being, which globally feels no force despite the fact that it is the effect and origin of contrary forces.

Let us assume that the original gravitational force F_u resolved into contrary forces F and R radiates away from the mass of the center a. As distance from the center a increases, attractive gravity F (or the attractive action of the original gravitational force F_u) is weakened by being spread across larger surface areas of concentric spheres of center a. The surface area of each three-dimensional sphere increases proportionally to the square of its radius, which is the distance from the center a, and so attractive gravity F falls (according to Newton's inverse square law of force) as the inverse square of distance:

$$3.18 \qquad F \propto 1/d^{2},$$

where $\propto$ is the sign of proportion and d is the distance from the center a.[16] If d becomes the radius ab of the greatest sphere of center a—that is, infinite—then the strength of attractive gravity F is zero.

---

16. Because the inverse power of distance depends on the two-dimensional surface of the limiting sphere, it is equal to 2.

The inverse relation holds for the contrary force of repulsive gravity R. If $R = 1/F$, we then obtain the following relation:

3.19
$$R \propto d^2,$$

stipulating the contrary of Newton's law, which we may call the *square law of force*. As the distance from the center a increases, repulsive gravity R (or repulsive action of the original gravitational force F_u) is strengthened by being spread across larger surface areas of concentric spheres of center a. This means that repulsive gravity R increases proportionally to the square of distance from the center a. If d is the radius of the greatest sphere of center a—that is, infinite—then the strength of repulsive gravity is infinite. In total, when d is infinite, the strengths of attractive gravity F and repulsive gravity R are, respectively:

3.20
$$F = 0 \quad \text{and} \quad R = \infty.$$

Because $F = R$ implies $F = F \times R$ and $R = R \times F$, we infer:

3.21
$$F = 0 \times \infty = 1 \quad \text{and} \quad R = \infty \times 0 = 1,$$

or, to put it in another way:

3.22
$$F_u = (F = R) = 0 \times \infty = \infty \times 0 = 1.$$

Formulae 3.21 and 3.22 suggest that the limiting-sphere b rotating on the limiting-circle b experiences the contrary forces F and R, or the synthetic, original gravitational force F_u resolved into contrary forces, as complex, indeterminate wholes having the totality of strengths and to which we assign the real number 1. Accordingly, seen from inside, the attractive gravity F is zero and its opposite repulsive gravity R is infinite. Seen from outside, on the other hand, the attractive gravity F is infinite and its repulsive gravity R is zero. We then conclude that maximally distant bodies b on the limiting-circle b and under the contrary, infinite actions of the complex, gravitational force F_u accelerate infinitely away from and toward the center a, and hence infinitely away from and toward each other at the same time. By accelerating infinitely away from each other (infinite repulsion), they infinitely expand the physical whole, its definite space-time and wavelength of light, into an unlimited Euclidean (or hyperbolic) line containing an infinity of dimensions, magnitudes, and things. By infinitely accelerating toward each other (infinite attraction), they infinitely contract or curve the physical whole, its definite space–time and wavelength of light, into a limiting point, a cosmic singularity, having nothing, no dimension, no magnitude, and constituting the common and universal contact point of everything within the unlimited, Euclidean line.[17]

The geometric expansion and contraction of the physical whole of unit radius *ab* can proceed logarithmically by the infinite multiplication and division of *ab* by the pair number 2 or 10, which we call the *change factor,* such that there is a one-to-one correspondence between the terms of its expansion and contraction, and each term of the geometric variation is a constant multiple of the preceding term:

17. Given the fact that the original gravitational force F_u emanating from the complex mass is a complex, indeterminate whole resolved into contrary forces F and R, the law of universal gravitation must reflect the force's complex nature. Accordingly, the complete law of universal gravitation must state that everything exerts both an attractive and a repulsive gravitational force on itself and on everything else. However, this law of complex, gravitational force contradicts our particular senses, which are powerless to perceive the *unity* of the opposite forces F and R and hence the complex nature of the gravitational force. The imperceptible unity of the forces F and R means that we experience the complex mass *as if* it were a simple mass being at one time the cause or the effect of a unique, simple force.

This simple nature of mass or gravitational force is expressed clearly by Newton's empirical law of universal gravitation, which states that everything exerts an attractive gravitational force on everything else. By revealing *uniquely* the attractive behavior of the gravitational force, it assumes that gravitational force has a simple nature admitting at one time a unique determination or contradictory determinations. However, what this empirical law describes is not the real gravitational force itself independent of our particular senses, but rather our partial and conflicting manner of perceiving the different determinations of the gravitational force, which is at one time according to the analytic principle of excluded third either attractive or repulsive. In this sense, Newton's empirical law of universal gravitation is incomplete, as it says something about the conflicting manner with which we experience the object but nothing about the real object itself independent of our analytic, particular experience.

3.23
$$ab \times 10^1, \quad ab \times 10^2, \quad ...ab \times 10^n, \ ... \ = \ \infty$$
$$ab \times 1/10^1, \quad ab \times 1/10^2, \quad ...ab \times 1/10^n,... = \ 0,$$

The simultaneous infinite expansion and infinite contraction of the physical infinite whole assigns to its motion a circular or vibrational form and to its different quantities, such as radius ab, curvature K, limiting-circumference b, force F_u, and speed v, the complex magnitude $1 = \infty \times 0$ numbered by the real 1. Thus, the geometric form of the physical infinite whole, the nature of its motion, and the magnitude of its quantities arise as an equilibrium state between opposing forces F and R or opposing actions of the same complex force F_u.[18] The balance between opposing nonuniform motions does not result in a uniquely static, spherical whole without motion (which is the Aristotelian definition of balance). Rather, this balance results in a spherical whole that is both at rest and in permanent rotation at the maximum speed $1 = \infty \times 0$.

---

18. If a particular observer experiences *uniquely* the expanding part of the physical, infinite whole, he or she concludes that the physical whole is indefinitely expanding under the *unique* force of repulsive gravity generating destructive asymmetries such as multiplicity without unity, distance without contact, and variation without constancy. Per contra, if a particular observer experiences *uniquely* the contracting part of the physical whole, he or she concludes that the physical whole is indefinitely contracting under the *unique* force of attractive gravity generating destructive imbalances such as unity without multiplicity, contact without distance, and constancy without variation. Each particular observer experiences at one time one and only one part of the physical whole, either the expanding part or the contracting part; this unique observable part has a unique sense and temporal order that causes the imbalance and collapse of its observable part. Per contra, the balanced, physical whole itself, defined as the sum total of all its parts, senses, and temporal orders, is free of temporal order by comprising the totality of temporal orders. Indeed, at any time the permanently rotating and vibrating physical infinite whole 1 both infinitely expands and infinitely contracts and neither infinitely expands nor infinitely contracts without absurdity.

Compression

An immanent property of the cosmic singularity, which we have identified with the real, physical whole's outer limiting-circle b, is the power of compressing everything into nothing. This power of maximum or infinite compression, which occurs naturally without constraint or effort, consists of compressing the physical whole's infinitely extended and infinite-dimensional Euclidean space-time containing an infinite number of parts of increasing mass within its cosmic singularity of zero extension.[19] These three kinds of maximum material, spatial, and temporal compression, involving maximum density of matter and energy, maximum temperature, and maximum speed, are necessary for materializing the highest unity, continuity, and uniformity of the physical whole and also for actualizing the continuous and immediate communication between its infinitely distant and diverse parts. These compressions allow the following:

- thinking, perceiving, counting, measuring, and traveling the physical whole's infinite distance at once (at one time or at zero time), with one concept, one number, and one action;
- affirming a universal and common proposition about an infinite multiplicity of particular things contained in the infinite extension of the physical whole independently of their particularities, which we call universal principle, law, or structure;
- verifying empirically that the jointly false universal propositions a = every x is a and a' = every x is not-a are jointly true and hence completely equivalent;

19. We define the cosmic singularity's infinite density as the ratio of any finite mass to zero volume. If the mass is that of the greatest body (the physical universe), whereas the volume is that of the smallest body (the cosmic singularity), then the cosmic singularity's material density is complex, relating the *greatest* mass to the *smallest* volume. On the other hand, if the mass and the volume are that of the smallest body, then the cosmic singularity's material density is simple, relating the *smallest* mass to the *smallest* volume. Because the infinitely dense cosmic singularity can support the entire weight of the physical whole without being broken, we conclude that it has infinite solidity and resistance.

- moving continuously and universally (everywhere, in all directions, at all times, under all conditions) at a maximum speed, within the universal community of the real, physical, infinite whole.[20]

20. In order to express geometrically the *complete* equality of the universal propositions a = every x is a and a' = every x is not-a (or every x is a')—that the jointly false universal propositions a and a' are jointly true, such that it is true that every x is a and every x is not-a or that every x is both a and not-a —we proceed as follows: We draw at once, by passing through the common intersection point u, the horizontal axis a, the equal and opposite vertical axis a', and the equidistant diagonal axis u of a square. The simple horizontal axis a represents the true, analytic universal proposition a, which affirms that every x of the horizontal axis a has the property a; the simple vertical axis a' represents the true, analytic universal proposition a', which affirms that every x of the vertical axis does not have the property a (or that every x of the vertical axis has the property a'). Now the synthetic unity (or product) of these opposite and equally true analytic universal propositions a and a' gives the synthetic universal proposition $u = (a = a')$ or $u = aa'$, which we represent by the diagonal axis u and which affirms that both universal propositions a and a' are true—that every x of the complex diagonal axis u is simultaneously a x of the horizontal axis a that has the property a and a x of the vertical axis a' that does not have the property a or a x of the vertical axis that has the property a':

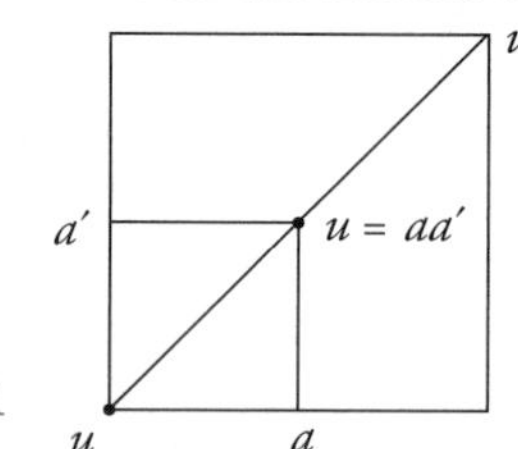

Fig. 1

We then define the synthetic universal proposition u as the proposition that both affirms and denies something of the object x universally without contradiction or paradox.

The diagonal axis u equally represents the impartial and indeterminate universal proposition u, whereby u is the synthetic unity or product of the opposite and equally false analytic universal propositions a and a': $u = [a' = (a')']$ or $u = a'(a')'$. Accordingly, the indeterminate proposition $u = a'(a')'$, which is equal to $(a + a')'$, affirms that neither (not-either) the universal prop-

Substance
The cosmic singularity of infinite density assigns substance—that is, independent and permanent existence—to the indefinitely varying

---

osition a is true nor the universal proposition a' is true, and hence that every object x has neither the property a nor the property a'. Indeed, every x of the equidistant diagonal axis u is neither a x of the horizontal axis a that has the property a nor a x of the vertical axis a' that does not have the property a (or a x of the vertical axis a' that has the property a'):

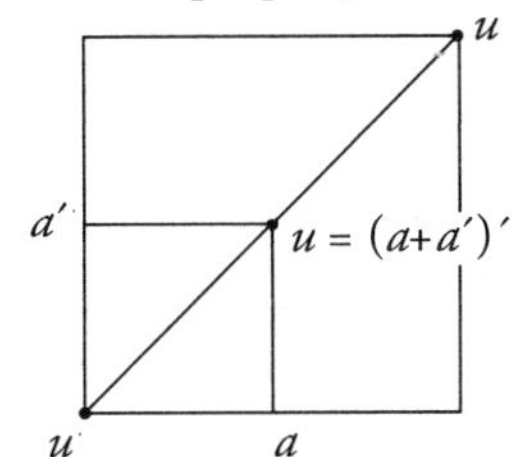

Fig. 2

We then define the impartial and indeterminate universal proposition u as the perfectly consistent and true proposition, which neither affirms nor denies something of the object x universally. Ultimately, we define the synthetic universal proposition $u = (a = a') = (aa')(a + a')'$, which is the synthetic unity of opposite and jointly true and false propositions a and a', as the proposition that both affirms and denies and neither affirms nor denies something of the object x universally without absurdity.

Because the object x is thought of as the complex, indeterminate universe, admitting contrary determinations and verifying synthetic principles of being, the synthetic universal proposition u is a perfectly true proposition that describes exactly the complex and indeterminate nature of the object x. In this sense, u is the proposition that enables us to know intellectually and with exactitude the object x existing in itself as a complex, indeterminate universe independently of our particular senses.

Taking this into consideration, we conclude that the geometric topos and geometric representation of the complex, indeterminate object x is the complex, indeterminate diagonal line u, which is both a *dividing line* dividing (resolving) the object x and its corresponding synthetic, universal proposition u into its opposite horizontal and vertical parts a and a' and a composition or *proportionality line* uniting these opposite parts.

Euclidean part. At the outer limiting-circumference b (cosmic singularity), the Euclidean finite part a_n varying indefinitely and successively relative to our particular senses becomes the ideal, spherical body of the physical, infinite whole 1 of radius equal to the real $1 = \infty \times 0$, which is a universal, permanent substance existing independently of

Another way of representing geometrically the complex, indeterminate object x and its corresponding synthetic universal proposition u is through the one-dimensional circle, which is both a dividing line resolving the object x and the universal proposition u into its opposite inner-flat (or concave) and outer-non-flat (or convex) parts a and a' and a composing line uniting these opposite parts:

$$u = (a\,a')\,(a + a')'$$

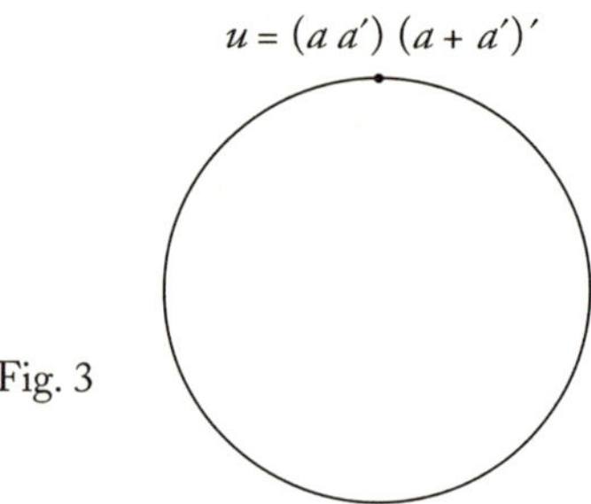

Fig. 3

Accordingly, if the inner flat (or concave) part of the circle represents the universal proposition a = every x is a and the outer non-flat (or convex) part of the circle represents the universal proposition a' = every x is not-a (or every x is a'), then the synthetic universal proposition $u = (a = a') = (aa')(a + a')'$, which is the synthetic unity or product of jointly true and jointly false propositions a and a' states the following: Every x of the complex and indeterminate circle u is both and neither a x of the flat (or concave) part that has the property a and a x of the non-flat (or non-concave) part that does not have the property a (or a x of the convex part that has the property a').

By means of the above geometric figures, we have shown how two conflicting analytic universal propositions, a and a', that cannot both be true relative to our analytic understanding are reconciled so that both are true without absurdity relative to our synthetic reason, regarded as the faculty of thinking of the object x as a complex, indeterminate universe receptive of contrary determinations.

our particular senses.[21] Because every contained part a_n is simultaneously the containing whole 1 in conformity with the whole–part equivalence principle, which stipulates the equality between the contained part a_n and the containing whole 1, we can define the containing whole 1 as the limit of an infinite number of wholes: the *whole of wholes.*

The property of being simultaneously a contained part and a containing whole we call *self-containment.* This property characterizes the spherical, real infinite whole 1 and all of its parts governed by the whole–part equivalence principle. If we call *conditioned effect* the contained part a_n at the inner center *a* and *unconditioned cause* the containing whole 1 on the outer limiting-circumference *b*, the property of self-containment then becomes the principle of self-causality, which is a principle of reflexive order. Because according to the principle of self-causality (spontaneous or immanent causality) every conditioned effect is simultaneously an unconditioned cause, there is no need for an external cause to explain and justify its existence.

Without the cosmic singularity to assign limit and completeness, constancy and certainty, substance and material unity to the physical, infinite whole, it disintegrates into an unlimited sequence of isolated parts (the individuals) deprived of unity and continuous motion. In fact, this is what happens when our analytic, particular senses perceive the physical, infinite whole 1 improperly as different from what it *is*—namely, *as if* it were an indefinitely accelerating finite part a_n without a cosmic singularity to realize its material unity, or, to put it in another way, with a cosmic singularity to realize its material unity either before or after the accelerating part a_n and hence *externally* to the accelerating

---

21. By *substance* (ουσία), we mean τό καθαυτό, what it *is*, or what exists in itself independently of the partiality and temporal order of our Euclidean, particular senses. Fundamental properties of substance are therefore *independence* and *permanence.* We have argued that the substance of the real, physical body, defined as the sum total of its infinite number of parts, is the spherical, infinite whole having as magnitude the unit distance *ab* equal to the real $1 = \infty \times 0$.

part a_n but never *simultaneously* with and hence *immanently* to the accelerating part a_n.

It follows that insofar as our analytic, particular senses experience the material unity of the physical whole *as if* it were an impossible unity realized by a *transcendent* cosmic singularity whose existence within the body of the physical whole involves the destruction of the physical whole, continuous and immediate motion between the parts of the physical whole is physically impossible. Only if our senses experience the unifying cosmic singularity as it *really is*—as the *transcendental* component of the physical whole, which is both external and immanent to the physical whole—will we be able to experience the material unity of the physical whole, that *all is one*, and be able to communicate immediately and universally with any or all of its infinitely diverse and distant parts.

We call infinite, synthetic, universal sensibility the ideal faculty of sensing the infinite multiplicity of isolated things as unity and one thing and the unlimited sequence of discontinuous parts as the limit and total sum of an infinite number of continuous parts: as an infinite whole regardless of whether the distance and difference between the parts are infinite.

But do we possess, as a finite brain composed of analytically behaving cells that operate at the level of low energy and low unity, such an infinite, synthetic, universal sensibility?

The Definite Solution to the Problem of Motion Resides in the Possession of the Ideal Faculty of Synthetic Universal Sensibility

Reason grasps the universal, whereas the senses grasp the particular.
—Aristotle, *Physics*

We have discovered that continuous motion from a to something entirely different b needs an equivalence link between a and b that will bridge their maximum difference and distance. Only if our synthetic, universal reason stipulates *a priori* an equality link between different things a and b can we replace a by its equal b, and thus effect the continuous passage from a to b. The founding principle of continuous motion along the unit distance $ab = 1$ is the *synthetic principle of equivalence*, which in its general form stipulates the equality of unequal things.

If the unequal things a and b are the *unlimited series* $a = a_n$ and the *limit* $b = 1$, then the equivalence principle takes the specific form of the finite–infinite equivalence principle, which stipulates the equality of the finite and the infinite—that every limit is an unlimited series and that every unlimited series is a limit. If the unequal things are the *part a_n* and the *whole* 1, then the equivalence principle takes the specific form of the whole–part equivalence principle, which stipulates the equality of the whole and the part—that every whole is a part and that every part is a whole.

However, the equality of unequal things that grounds continuous motion from *a* to *b* is judged by the *analytic principle of contradiction* (which is a principle of our imperfect faculties of finite particular sensibility and analytic understanding) to be absurd or impossible. Indeed, if the unit distance *ab* is a Euclidean distance, where the unequal things $a < b$ are unequal and determinate, then how is it possible that unequal things are equal and indeterminate? It is clear that the equality of unequal things and continuous motion grounded in this equality are necessarily an impossibility for the Euclidean unit distance *ab*.

Given the contradiction between the Euclidean, analytic ontology of the unit distance *ab* and its property of continuous motion requiring a different ontology, we choose to proceed directly to the solution of the contradiction, instead of rescuing our Euclidean, analytic perception of things. To do this, we construct the rational and physical foundation of continuous motion. For this construction, we replace the *analytic* theory of infinite convergent series, where the equality of unequal things (of the unlimited series a_n and the limit 1) is an appearance dissimilating and regulating a permanent inequality between them, with the *synthetic* theory of infinite convergent series, where the equality of unequal things is a substance; in fact, this equality is the constitutive principle of the unit distance *ab*. If the unit distance *ab* has as its constitutive principle the synthetic principle of equivalence, which affirms that unequal things are equal and indeterminate, then the geometry of the unit distance *ab* and hence of the extended, physical body is not Euclidean but rather spherical. In this case, the different and unequal things $a < b$ of the unit distance *ab*, instead of occurring in Euclidean space, occur on the complex and indeterminate limiting circle of the limiting sphere of diameter $ab = 1 = \infty \times 0$. This limiting circle is an *infinitely curved straight line* whose complex unit curvature is the product of zero and infinite magnitudes. Thus, the maximally different and distant things *a* and *b*, represented geometrically by the diametrically different and unequal points $a < b$ of the inner flat (or concave) part of the limiting circle, are maximally united by the limting circle's outer infinitely curved part—the cosmic singularity part (see fig. 3.6). We have then, the equality of unequal points $(a = b)(a < b)$ occurring on the complex limiting circle of the limiting sphere.

Thus, it is not sufficient to postulate, through our ideal faculty of synthetic, universal reason, the synthetic equivalence principle as the founding principle of continuous motion in the unit distance *ab*. We must also assume that the unequal points $a < b$ of the unit distance *ab* are situated on the complex, indeterminate limiting circle of the limiting sphere of diameter $ab = 1 = \infty \times 0$. Without the unifying cosmic singularity of the limiting circle, continuous, immediate motion between *a* and *b* would be physically impossible; and the synthetic principle of equivalence grounding continuous, immediate motion would be a purely intellectual principle of our ideal faculty of infinite, synthetic, universal reason, having no substantial, physical, or objective existence whatsoever.

Having given the logical and geometric solutions to the problem of continuous motion, we now proceed to determine the empirical solution to the problem of motion, which is simultaneously the definite solution completing the two preceding solutions. By *definite solution* we mean the faculty to experience effectively the continuous motion from *a* to *b* despite their Euclidean infinite separation and distance. This experience is equivalent to traveling, counting, and perceiving the infinite of extension at once—that is, in a finite time or in zero time. Because

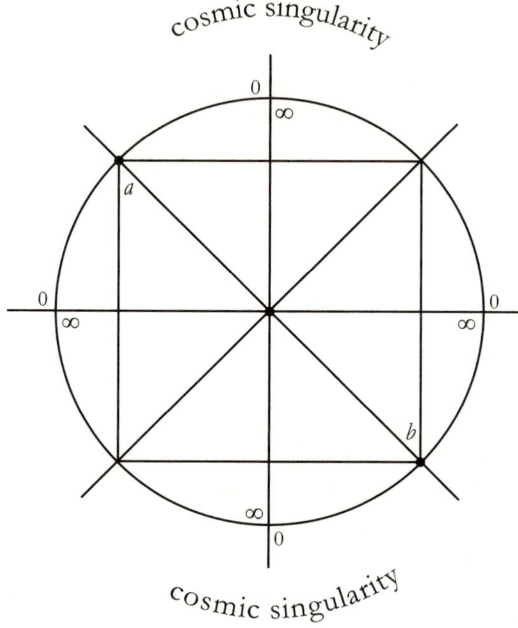

Figure 3.6. The maximally different and distant things *a* and *b* represented geometrically by the diametrically opposite and unequal points *a* < *b* of the inner flat (or concave) part of the limiting circle are maximally united by the limiting circle's outer infinitely curved part—the cosmic singularity. On the whole, we have the equality and indetermination of any two unequal points *a* and *b* of the complex, indeterminate limiting circle: ($a = b$)($a < b$).

everything (the Euclidean infinite distance *ab* containing everything) is compressed into its cosmic singularity of zero radius, it is sufficient to touch this cosmic singularity with our finite body (brain) in order to travel, count, measure, and perceive the infinite distance at once.

Let us call *universal sense* the particular sense elevated to the level of highest unity. The organ of the particular sense is our finite, individual brain; the organ of universal sense is the infinite, universal brain, which is nothing more than our finite, individual brain elevated to the level of highest unity. Because this universal and highest unity is actualized by the cosmic singularity of the limiting circle, it occurs on the outer, convex part of the limiting circle.

We call *infinite, synthetic, universal sensibility* the ideal faculty of sensing the physical object as it *really is,* as an infinite whole, as an infinite multiplicity of parts with universal unity and continuous, immediate motion.

Given the fact that we have *actually* no ideal faculty of infinite, synthetic, universal sensibility for experiencing the universal unity and immediate motion between any pair of infinitely distant parts *a* and *b* of the infinite multiplicity of parts, can we conclude that a complete solution to the problem of motion requiring an empirical solution is impossible?

We argue that our lack of an adequate faculty for experiencing the universal unity of *a* and *b* and their continuous, immediate communication is an accident or an appearance. In reality, the ideal faculty of infinite, synthetic, universal sensibility is one of the four fundamental cognitive faculties that we necessarily possess as a knowing subject.

Since the time of the ancient Greeks, it was known that we have two distinct ways of knowing the real, physical object:[22]

---

22. Aristotle remarks in his *Physics* I (6), 189a, 5, that the universal is known according to universal reason, whereas the particular or individual is known according to particular sensation (τό μέν γάρ καθόλου κατά τόν λόγον γνώριμον τό δέ καθ' ἕκαστον κατά τήν αἴσθησιν).

1. The rational or universal way of knowing the real, physical object consists of using our universal reason (λόγος νοῦς) for thinking the real, physical object completely from all sides (spherical cognition), as an infinite whole of magnitude $1 = \infty \times 0$ numbered by the real 1 and defined as the sum total of an infinite number of parts, as a timeless universe or maximum admitting at any time contrary determinations (senses, sides, parts), and as governed by synthetic principles of being—for example, by the principle of equivalence and zero temporal order. The geometry of the physical, infinite whole is spherical, comprising the totality of curvatures and forms; it thus is an unlimited Euclidean (or hyperbolic) plane of zero curvature when seen from inside space-time and a spherical limiting point of infinite curvature when seen from outside space-time. Contrary determinations assign balance, proportionality, permanent life, and continuous rotational motion to the physical, infinite whole 1.

2. The empirical or particular way of knowing the real, physical object consists of using our particular senses for perceiving the real, physical object uniquely from inside (Euclidean cognition) as an incomplete, Euclidean part of magnitude a_n, as the partial sum of an infinite series of discontinuous parts, as an individual admitting at one time a unique sense or contradictory senses, and as governed by analytic principles of being—by the principle of inequality and temporal order; for example, if time is space with a unique sense, and anything extended having a unique sense is time-conditioned, then necessarily the sense of its time is its destruction as a time-conditioned thing. It follows that insofar as the Euclidean extended part a_n has a unique sense, it is necessarily time-conditioned, its arrow pointing at the destruction of the Euclidean part a_n either by being infinitely stretched into a cold plane or by being infinitely compressed into a hot point.

Let the term *aa* designate our imperfect faculty of particular sensibility, which is Euclidean, finite, and analytic. This imperfect faculty (*aa*) perceives the physical object as a time-conditioned, finite part a_n (the individual) having no unity between its discontinuous parts. To put it another way, this faculty (*aa*) perceives the multiplicity of the physical object discontinuously and successively.

Let the term *bb* designate our ideal faculty of universal reason, which is spherical, infinite, and synthetic. This ideal faculty (*bb*) thinks of the physical object as a timeless, infinite whole 1 (the universe) having the highest unity between its continuous parts.

Universal reason is equally the faculty of synthetic ideas and synthetic principles that govern the physical object thought of as a complex, infinite whole. Because as knowing subjects we possess both faculties *aa* and *bb* of knowing the physical object, these faculties are simultaneous and equal: *aa* = *bb*.

If we interchange the terms of the equivalence *aa* = *bb*, we then obtain two additional faculties of knowing the physical object—*ab* and *ba*. The term *ab* designates the imperfect faculty of particular reason, which we call understanding. *Understanding* is the intellectual faculty that translates the perceptions of our particular senses into analytic concepts and analytic principles. In this sense, analytic understanding is the faculty of thinking the physical object's infinite multiplicity successively without unity, with partial unity, or with low unity between its parts. The converse term, *ba*, designates the ideal faculty of universal sensibility, which is the sensuous counterpart of our universal reason. Universal sensibility is in turn divided into *transcendental imagination* (the faculty that intuits formally or intellectually the infinite wholeness of the physical object) and proper *universal sensibility* (the faculty that intuits materially or sensuously the infinite wholeness of the physical object). Proper universal sensibility is the faculty that perceives the infinite multiplicity of the physical object at once as an infinite whole with highest unity between its parts. In this sense, universal sensibility must have the power to experience the physical object's highest unity— its outer cosmic singularity unifying all things. In total, the knowing subject possesses the four fundamental faculties *aa, bb, ab,* and *ba* for knowing the physical object. The low faculties *aa* and *ab* determine the knowing subject's imperfect individual mind; the high faculties *bb* and *ba* determine the knowing subject's ideal universal mind.

We have shown that the ideal faculty of infinite, synthetic, universal sensibility is the necessary consequence of our primitive assumption that we possess simultaneously two fundamental ways of knowing the physical object: first, the rational–universal way, through which we know the physical object as a physical, infinite whole 1 with highest unity between its parts and verifying synthetic principles of being; and second, the empirical–particular way, through which we know the physical object as an incomplete finite part a_n, with low unity between its parts and verifying analytic principles of being. The relevant question therefore is not whether the infinite, universal sensibility exists as the sensuous counterpart of our infinite, universal reason, for it exists and it is real, at least in principle, intellectually and potentially, but rather how it is possible in act and in experience.[23]

23. The ideal faculty of sensing the universal, which we call infinite, synthetic universal sensibility, corresponds to Aristotle's concept of τού καθόλου αίσθησις—the sensation of the universal (see Aristotle's *Analytics* II, 13–100a, 16). However, Aristotle makes a metaphoric use of this powerful synthetic concept by considering it as an improper activity of universal reason, which senses *intellectually* the universal, and not as a proper activity of universal sense, which senses *materially* or *sensuously* the universal. We go a step further and demand a literal and proper use of τού καθόλου αίσθησις if we want to provide a definite solution to the problem of continuous motion and universal unity between discontinuous and isolated parts of the physical whole. The proper use of τού καθόλου αίσθησις has as its ultimate goal the activation of our ideal faculty of infinite synthetic, universal sensibility so that we sense *effectively* the universal unity of things—that is, the infinite sequence of things as an infinite whole in which continuous and immediate communication is real and complete.

• 4 •

Physics of Universal Perception

Topology of the Individual Brain and the Universal Brain in the Physical Universe

What is the first principle of the physical universe
written on its luminous surface?

*T*he real, physical universe, numbered by the real 1, is the limit and total sum of an infinite series of particular bodies of increasing radii. We consider the real, physical universe to be the perfect or maximum body—that is, the greatest and fastest body according to quantitative order, the highest body according to position in ontological and logical order, the earliest and latest or the *first* and *final body* according to position in chronological order, and the *first* and *final cause* according to position in causal order. Because the real, physical universe is a maximum quantity, it necessarily has a cosmic singularity ending its infinite magnitude given by its infinite series of parts of positive magnitude. By ending its infinite magnitude, the physical universe has no magnitude greater than its infinite magnitude, which then becomes the *greatest* magnitude. The synthetic product of infinite and zero magnitudes gives the complex, indeterminate magnitude $1 = \infty \times 0$, which is the real number of the infinite whole or of the infinite in act—the substance of the real, physical universe.

As the body of all bodies, the real, physical universe numbered by the real 1 is, according to the synthetic principle of reflexive order, both a contained part and a containing whole and hence is self-contained and self-ordered or spontaneously ordered:

4.1 $$1 = (1 < 1)(1 < 1) \rightarrow 1 < 1.$$

Because the physical universe is self-ordered, it is at one and the same time earlier than itself (earlier than the earliest) and later than itself (later than the latest) according to the chronological order.

The physical universe is simultaneously smaller than itself (smaller than the smallest) and greater than itself (greater than the greatest) according to the quantitative order.

As a containing whole, the physical universe does not have the property of its parts—that is, of a corporeal existence. As a contained part, the physical universe has the property of its parts—that is, of a corporeal existence. Ultimately, the complex, physical universe has both a corporeal and an incorporeal existence; it is both a body and the end of a body without absurdity. The physical universe verifies, therefore, the synthetic principle of equivalence, which stipulates the equality of unequal things, of the corporeal and the incorporeal. On the other hand, grounded in the synthetic principle of included third, the same physical universe has neither a corporeal existence nor an incorporeal existence, and hence it is an indeterminate and impartial or neutral being.

By being earlier and later, greater and smaller than itself, the maximum and limited physical universe is simultaneously unlimited in time and space. The physical universe therefore verifies the finite–infinite equivalence principle, which stipulates that every physical body is both limited and unlimited, closed and open, spherical and flat with respect to space and time. The synthesis of the spherical and flat determines the nature of the physical universe's sphericity, which is complex, indeterminate, and impartial. Indeed, in our intuitive one-dimensional description of the physical, infinite whole, we have asserted that this infinite whole is a limiting circle produced by the intersection of the limiting sphere of zero radius with the unlimited Euclidean plane of infinite radius passing through the center of the limiting sphere (see fig. 3.2, p. 60).

Thus, an ideal body b on the limiting-circle b of the sphere has the power and freedom to do at one time anything or everything—namely, to recede away from the starting point, the center a, in a straight (or hyperbolic) line at an increasing speed and under the unique and imbalanced action of repulsive gravity and to rotate about the center a in a circular line at a maximum speed and under the balanced action of contrary forces—of repulsive and attractive gravity. The same ideal, moving body b may also realize the totality of cases at the same time if it is divided into contrary moving bodies b_1, and b_2; the first body b_1 recedes away from the center a in a straight (or hyperbolic) line at an increasing speed and under the unique force of repulsive gravity; the second body b_2 rotates about the center a in a circular line at a maximum speed and under the contrary (equal and opposite) repulsive and attractive forces of gravity. To accelerate away from the center a without returning is equivalent to approaching indefinitely the limiting-circle b without ever attaining the limiting-circle b. In both cases, the body recedes in a boundless Euclidean universe at an indefinitely increasing speed. On the other hand, to rotate about the center a at a maximum speed is equivalent to traveling instantaneously from one point to another point of the straight line by passing via the cosmic singularity beyond the straight line at the maximum speed $1 = \infty \times 0$.

Now, the impossibility of deciding which of the above geometric forms—the circular or the rectilinear (relative to the one-dimensional universe), the spherical or the flat (relative to the two-dimensional universe)—is the real form of the physical universe does not indicate our incomplete knowledge and inexact description of the universe. On the contrary, this impossibility reveals our complete and exact description of the universe, which, in opposition to the analytic-Aristotelian universe, is a complex and indeterminate universe verifying synthetic principles of being.

We therefore must distinguish between two kinds of definitions of the real, physical universe. These definitions are the analytic-Aristotelian definition, according to which the universe is a simple and determinate sphere admitting no spatial contrariety, and the synthetic–non-Aristotelian definition, according to which the universe is a complex and indeterminate sphere receptive of spatial contrariety. For example, the radius of the Aristotelian sphere is simple and determinate, and hence (in conformity with the analytic principle of excluded third) it is at one time either limited or unlimited. If the radius is limited, then the Aristotelian sphere is reduced to a closed sphere; if the radius is unlimited, then the Aristotelian sphere is reduced to an open, unlimited plane, which is Euclidean (or hyperbolic). In both cases, the Aristotelian sphere verifies analytic principles of being. Per contra, the radius of the non-Aristotelian sphere is complex and indeterminate and hence (in conforming to the finite–infinite equivalence principle) is both limited and unlimited, whereas (in conformity with the synthetic principle of included third) it is neither limited nor unlimited. Accordingly, from inside, the non-Aristotelian sphere has an infinite radius and therefore is a Euclidean or hyperbolic plane; from outside, the non-Aristotelian sphere has a zero radius and hence is a limiting point. Combined, these contrary radii give, as we have argued, a complex, indeterminate, or impartial radius equal to the real $1 = \infty \times 0$.

The first, analytic definition of the real, physical universe is an improper or false definition; actually, it defines the simple (either closed or open) sphere of the individual, which is merely a determinate part of the physical universe but not the physical universe itself. If the physical universe is anything having contrary determinations, then the true, proper, and exact definition of the physical universe is the synthetic–non-Aristotelian definition. Spatial contrariety is therefore neither an accidental property (Aristotelian thesis) nor an apparent property (Kantian thesis) of the physical universe, but rather its immanent nature. Far from being a sign of our incomplete knowledge of the physical universe, spatial contrariety determines our all-comprehensive knowledge of the complex (open–closed) sphere of the universe numbered by the real 1.[1]

1. One of the unanswerable questions of cosmology concerns the geometry of the real, physical universe: whether it is closed or open, limited or unlimited, with respect to space and time. Since the dawn of natural philosophy the physical universe, identified with the Divine Being, has been thought of as a neutral and indeterminate being that is neither limited nor unlimited in its space and time dimensions (for example, see Ionian Greek philosophy and Indian philosophy). However, this neutral and indeterminate physical universe has been understood improperly (falsely) by our empirical faculty of analytic understanding, which assumed from the time of Aristotle (see *Physics*, *Metaphysics*) the simple, determinate nature of the being: that the being is an individual.

We may call *Aristotelian* or *analytic* the ontological doctrine affirming that anything existing in itself and real is a simple, determinate individual, admitting at one instant a unique determination or contradictory determinations and verifying analytic principles of being. If anything existing in itself and real is simple and determinate verifying the analytic principle of excluded third, then insofar as the physical universe exists in itself and is real, it is either limited or unlimited with respect to its space and time dimensions. However, our synthetic, universal reason shows us that the real, physical universe represented by the limiting-sphere b is both limited and unlimited and neither limited nor unlimited with respect to its space and time. We conclude, then, from the analytic ontological point of view, that the real, physical universe is self-contradictory and therefore nonexisting and unreal. We also conclude, from the analytic epistemological point of view, that even if the physical universe exists and is real, we cannot know and describe its immanent nature, which is either limited or unlimited with respect to its space and time dimensions.

The first conclusion of analytic ontology leads to *ontological nihilism* (*irrealism*), according to which the real, physical universe does not exist, or, if it exists, its fate is its nonexistence, or its ephemeral, apparent, or illusory existence. The second conclusion of analytic ontology leads to *epistemological nihilism*, called *skepticism*. This position states that even if the physical universe exists and is

In fig. 3.3 (p. 61), let the inner center a of the limiting circle, which we call *here and now*, be the intersection of a three-dimensional space represented by the horizontal axis xx and a one-dimensional time represented by the vertical axis yy. The proportionality of space and time is expressed by the square's diagonal axis zz, which is the path of light.

real, it remains unknown and inaccessible to us. For example, the real, physical universe (numbered by the real 1) is inaccessible by the indefinitely approaching a_n because the limit 1 and the unlimited series a_n are assumed to be unequal and analytically separated (see the analytic theory of infinite convergent series (pp. 23–33). If the real $1 = \infty \times 0$ is the definitive answer to the cosmological question of the geometry of the physical universe, then for the indefinitely approaching a_n, assuming an analytic ontology of the real 1, the complex, indeterminate nature of the real 1 is incomprehensible or absurd. In fact, the real 1 (as the simple answer to a_n's cosmological question) is impossible; and even if the simple and determinate answer exists, it is unknown by a_n, a fact leaving the cosmological question about the geometry of the physical universe unanswerable.

The impossibility of knowing the real nature of the physical universe, assumed unconsciously and improperly as being simple and determinate, gave rise to different doctrines of skepticism during the growth of science. For example, it engendered Nicolas of Cusa's *Docte Ignorantia*, which affirms the impossibility of formulating a simple, univocal representation of the universe; Kant's transcendental idealism, which affirms the impossibility of knowing the universe as a simple whole existing in itself, which is either finite or infinite; and Gödel's incompleteness theorem of truth, which affirms the impossibility of stating a simple proposition about the universe that is either true or false, either proved or disproved.

In order to escape the trap of ontological and epistemological nihilism generated by an analytic ontology of the being that contradicts the being, we replace this analytic ontology by its negation. We call *synthetic* or *Platonic* the ontological doctrine affirming that anything existing in itself and real is a complex, indeterminate universe, having at one instant contrary properties and verifying synthetic principles of being. Thus, with respect to its space and time dimensions, the real, physical universe is (in conformity with the finite–infinite equivalence principle) both limited and unlimited; and in conformity with the

Let us next assign to anything located at the inner center a, regarded as the here and now, the intrinsic properties of being (1) low, heavy, and cold; (2) at rest (ground state) with a minimum or indefinitely increasing speed v_n and in a flat space-time of zero force; and (3) an *individual* or incomplete, finite part with sense and time (succession) and verifying analytic principles of being—say, the principles of reflexive

synthetic principle of included third, the real, physical universe is neither limited nor unlimited. Synthetic ontology leads us directly to *realism*, which affirms the continuous existence of the physical universe, whose substance is the complex, infinite whole. It also leads us to *gnosticism*, which affirms the knowledge and accessibility of the real, physical universe. For example, the real $1 = \infty \times 0$ (regarded both as the radius and curvature of the real universe and as the complete and exact answer to the cosmological question of the geometry of the real universe) is accessible and comprehensible by the approaching a_n, assuming a synthetic ontology of the real 1 (see the synthetic theory of infinite convergent series, pp. 37–46). Because of the complex and indeterminate nature of the real, physical universe, the true and complete answer to the question of the geometry of the universe (is the universe limited or unlimited with respect to space and time?) does not require the analytic choice of one of the alternatives, which is a false or incomplete answer. Rather, this question requires their mutual acceptance and transcendence generating the balanced state of a complex, indeterminate answer, which is itself the true and exact answer relative to the complex, indeterminate universe receptive of spatial contrariety.

We present the conversion of the analytic, ontological argument (leaving unanswered or incompletely answered the cosmological question about the geometry of the physical universe) into the synthetic, ontological argument (giving a complete answer to this cosmological question) according to the following formal terms.

Let us postulate the propositions:

a = the physical universe is a real being (a being existing in itself)

b = the physical universe is either finite or infinite

Assuming that the propositions a and b verify uniquely deductive rules of *modus ponens* and *modus tollens* such as $a < b = b' < a'$, we affirm the following: If the physical universe is a real being, then the physical universe is either finite or infinite. However, our synthetic, universal reason shows that it is false identity (analytic identity), excluded third, contradiction, inequality, and temporal order. As a simple, determinate quantity, the individual has at one time a unique magnitude or contradictory magnitudes and at different times successive magnitudes. The individual is therefore a finite variable that changes successively in time, and an object of our

that the physical universe is either finite or infinite; thus, it is false that the physical universe is a real being. We conclude, then, that the physical universe is unreal, an illusion, or, if we want to follow Kant's conclusion, that the physical universe is a representation, an appearance, having an inferior (secondary) ontological status (see *Critique of Pure Reason,* Transcendental Dialectic Book II, chapter 2, The Antinomy of Pure Reason, section VII, Critical Solution of the Cosmological Problem).

This skeptical thesis about the reality of the physical universe is reversed if we change by postulation the content of the proposition *b*:

Let us then stipulate the following propositions:

a = the physical universe is a real being (a being existing in itself)

b = the physical universe is is both finite and infinite and is neither finite nor infinite

Assuming that the propositions *a* and *b* verify the deductive rule of *modus ponens* and its converse, the inductive rule of *conversion* such as $a < b = b < a$, we affirm the following: If the physical universe is a real being, then the physical universe is both finite and infinite and is neither finite nor infinite. Because our ideal faculty of synthetic reason shows that the physical universe is both finite and infinite and is neither finite nor infinite, it follows that the physical universe is a real being.

In line with these considerations, we are led to the following conclusions about spatial contrariety: The assignment of contrary properties to the real being may determine or not determine the complete reality of the being depending on whether we assume a synthetic or an analytic ontology of the being. If we assume, through our analytic understanding, an analytic ontology of the being, that the real being is an *individual,* then spatial contrariety is an accidental, apparent, or absurd property of the being determining its unreality or incomplete reality and incomprehension. Per contra, if we assume, through our synthetic, universal reason, a synthetic ontology of the being, that the being is a *universe,* then spatial contrariety is a substantial, real, or consistent property of the being determining its complete reality and comprehension. In this latter particular sense—of our cognitive faculties of particular sensibility and analytic understanding.

Now, let us assign to anything located on the outer limiting-circumference b, which we regard as the farthest and earliest (or latest) space-time, the intrinsic properties of being (1) high, light, and hot; (2) in continuous rotation with a maximum speed equal to the real $1 = \infty \times 0$; and (3) a *universe* or a complete, infinite whole free of sense and time (succession), and verifying synthetic principles of being—say, the principles of reflexive order, included third, equivalence (irreflexive identity), and zero temporal order (zero succession). As a complex, indeterminate quantity, the universe equal to itself and 1 has at any time contrary magnitudes—infinite and zero magnitudes. It is therefore a universal constant, an *ideal* and *first quantity*, numbered by the real $1 = \infty \times 0$, which varies at once regardless of time (for example, ∞ is the limit of an infinity of increasing magnitudes and 0 is the limit of an infinity of decreasing magnitudes), and computes or measures any maximum quantity of the physical universe regardless of its kind.

Finally, a Euclidean, particular observer at the center a perceives the maximally distant points a and b, where a is the inner center and b is the outer limiting circumference of the universe, as maximally unequal and determinate points occurring in the inner, Euclidean part of the universe.

---

case, any complete and exact description of the real being must be complex and indeterminate, verifying synthetic principles of being; if the complete and exact description of the real being is simple (or univocal) and determinate, it is forced to produce absurdities.

Finally, it is not sufficient to have a complete and exact description of the real, physical universe through our ideal faculty of synthetic, universal reason. We must eventually obtain its spherical perception—that is, the faculty to perceive the physical universe universally and symmetrically from any or all sides—and hence to experience its contrary properties: namely, its infinitude and finitude, the open and closed parts of its complex and indeterminate sphere of radius $1 = \infty \times 0$.

The finiteness of our particular senses
determines the finite speed of sensible light,
which determines our partial experience of the universe.

Let us assume that light emitted from the limiting-circumference or limiting-point b travels one unit of space in one unit of time at the finite, sensible (observable) speed $c = 3 \times 10^8$ m/s, which we take as unit of maximum speed equal to 1. If space is represented by the horizontal axis xx and time is represented by the vertical axis yy, then the same light $c = 1$ that travels at the same time equal amounts of space and time along the horizontal and vertical axes xx and yy necessarily travels along the diagonal axis zz of a square whose opposite and equal sides are the axes xx and yy (see fig. 4.1). The diagonal axis zz is therefore an axis of proportionality expressing the synthetic unity and equality of opposite parts, say of space and time. Conversely, when light emitted from the limit b travels along the square's diagonal axis zz at the finite, sensible speed $c = 1$, the same light travels at the same time equal amounts of space and time along the horizontal and vertical axes xx and yy. Light traveling along the vertical axis yy (that is, uniquely through time) is light traveling a zero distance in one unit of time (finite time) or one unit of distance (finite distance) in infinite time at zero speed. Light traveling along the horizontal axis xx (that is, uniquely through space) is light traveling one unit of distance (finite distance) in zero time or an infinite distance in one unit of time (finite time) at infinite speed. We then call *ideal* or *real light* the physical light emitted from the limiting-circumference b, which travels along the complex diagonal axis zz at the maximum speed $c = 1$ resolved relative to contrary axes yy and xx into zero and infinite speeds, and to which we assign the real number 1 defined as the product, or ratio of infinite and zero magnitudes: $c = 1 = 0 \times \infty$, $c = 1 = \infty/0$ (see fig. 4.1). Thus, the real 1 is a constant, universal numeric magnitude measuring different quantities of the physical universe's greatest body, such as radius, curvature, limiting circumference, and motion; the real 1 is equally a principle of synthetic unity and contact between different parts (for example, between space and time, between zero and infinite

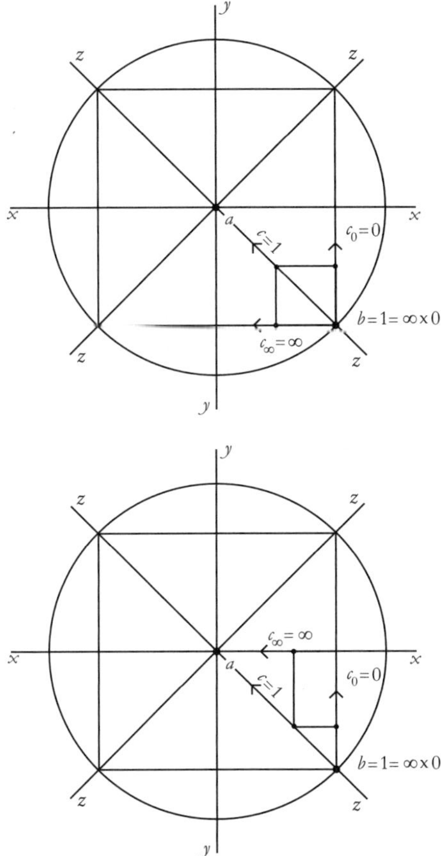

Figure 4.1. Two ways of showing the complex and indeterminate nature of light emitted from the physical universe's limiting-point b (which is the farthest and earliest point) and traveling along the complex and indeterminate diagonal axis zz. Because the complex diagonal axis zz can be divided (resolved) into contrary vertical and horizontal axes yy and xx, light traveling along the complex diagonal axis zz at the maximum speed $c = 1$ is light divided (resolved) into contrary lights c_0, and c_∞, traveling along contrary axes yy and xx at zero and infinite speeds and hence along the totality of axes at all speeds: $(c = 1)_{zz} = (c_0 = 0)_{yy} (c_\infty = \infty)_{xx}$.

magnitudes). In this sense, the natural place of the real 1 is also (besides the limiting-circle b) the square's diagonal axis zz thought of as a complex line uniting opposite parts. Finally, we can divide (resolve) light traveling along the complex diagonal axis zz at the maximum speed $c = 1$ into different lights c_0 and c_∞ traveling along the vertical and horizontal axes yy and xx at zero and infinite speeds (see fig. 4.1):

4.2
$$(c = 1)_{zz} = (c_0 = 0)_{yy} (c_\infty = \infty)_{xx}.$$

If we consider light c_0 traveling along the vertical axis yy at zero speed as the minimum limit of an infinite number of speeds c less than 1, and if we consider light c_∞ traveling along the horizontal axis xx at infinite speed as the maximum limit of an infinite number of speeds c greater than 1, then light traveling along the complex diagonal axis zz at the maximum speed $c = 1$ is divided into an infinite number of lights traveling along the totality of axes zz, yy, xx and at the totality of speeds c equal to 1, smaller than 1, and greater than 1:

4.3
$$(c = 1)_{zz} = (c < 1)_{yy} (1 < c)_{xx}.^2$$

---

2. Because the diagonal axis zz is the complex line, which is both vertical and horizontal, and the indeterminate line, which is neither vertical nor horizontal, the real number 1 on the diagonal axis zz has its synthetic and indeterminate properties. This means that the real number 1 is, similar to its diagonal axis zz, both a *dividing* and a *unifying* number. The real number 1 divides a given unit quantity into small and great—for example, it divides the unit quantity $c = 1$ into less than 1 and greater than 1—and at the same time unites or conjoins the inequalities or variations to obtain the complex unit quantity: $(c = 1) = (c < 1)(1 < c)$. We may likewise say that the real number 1 divides a unit quantity of a given kind—for example, $c = 1$ into zero and infinite magnitudes—and at the same time unites or conjoins these different magnitudes to obtain the complex unit quantity $c = 1 = 0 \times \infty$.

The diagonal axis zz is not only a complex and indeterminate line dividing the thing or its property into opposite parts and at the same time uniting them; it is also a line of constancy and balance—that is, the topos where anything, which is both a and a' and neither a nor a' relative to the contrary parts xx and

Because light's wavelength λ is the inverse of its speed c such as λ = $1/c$, whatever holds true for light's speed, the inverse holds true for light's wavelength. Accordingly, light that travels along the vertical axis yy travels at zero speed or with infinite wavelength, whereas light that travels along the horizontal axis xx travels at infinite speed or with zero wavelength.

On the basis of formulae 4.2 and 4.3, we can ultimately call *ideal* or *real light* the physical light that travels along all axes at all senses (forward and backward), at all speeds (or with all wavelengths), at all frequencies, and verifies synthetic principles of being. Because real, physical light is emitted from the limiting-point b, which we regard as the farthest and earliest point, we can qualify it as primordial light or *first light*. This first light carries information about the entire physical universe—its inner and outer parts, its early and late time, its past and future. However, our finite, particular senses (retinal cells) detect only that kind of light whose transmission speed c is 3×10^8 m/s and finite wavelength λ varies between 10^{-6} m (red light) and 10^{-7} m (blue light). This finite, optical portion of incident physical light we call *sensible* (optical) *light*. Although optical light is sensible light, sensible light is larger than optical light as it includes the part of physical light that is observed indirectly through artifacts.

Now, sensible light of finite wavelength λ that travels uniquely along the diagonal axis zz in a unique forward sense and at the unique, finite speed 3×10^8 m/s equal to 1 is a Euclidean, simple, determinate light that shows the universe and any part of the universe as *if* it were a simple individual. For example, sensible light shows the diagonal axis zz *as if* it were a simple, undivided line that cannot be resolved into contrary axes yy and xx and the sensible speed $c = 3 \times 10^8$ m/s = 1 *as if*

---

yy of the diagonal axis zz, is a balanced and permanent universe—a real infinite whole 1 verifying synthetic principles of being. Plato considered the diagonal line and the circular line to have the same synthetic properties and founding principles and thus as being equivalent.

it were a simple, undivided speed that cannot be resolved into contrary speeds relative to contrary axes yy and xx (see fig. 4.2).

Because sensible light at the Euclidean center a reveals to our backward, particular perception only the inner, past part of the physical universe and to our forward, particular action only the inner, future part of the physical universe, the center a is the origin of a double cone of sensible light—the light cone of backward past and forward future experienced successively because of the finite speed of sensible light. Thus, Euclidean, sensible light of unique, finite speed $c = 1$ reveals the multiplicity of the universe successively, as an impossible unity where different things—passive past and active future—are for example consecutive and unequal individuals.

Per contra, real light, which is the entire spherical light, transmitted in all senses (forward and backward) along the complex, diagonal axis zz at the maximum speed $c = 1 = 0 \times \infty$, or which is the same along all axes at all speeds, reveals the multiplicity of the universe simultaneously, as a necessary cosmic unity where different parts of the universe—for example, backward past and forward future—are contemporaneous and equal wholes (universes).

We can then consider the Euclidean center a as the origin of a double illusion—the perceptual and onto-epistemological illusion from which our imperfect cognitive faculties of particular sensibility and analytic understanding suffer. The perceptual illusion consists of experiencing through our particular sensibility the complex, spherical light of the physical whole as different from what it *is*: namely *as if* it were a simple, Euclidean light that shows the multiplicity a_n of the physical whole 1 to vary successively and indefinitely in time in conformity with analytic principles of organization. The onto-epistemological illusion, on the other hand, consists of thinking through our analytic understanding this indefinitely varying multiplicity a_n *as if* it were the true and real, physical whole, and the analytic principles that govern a_n *as if* they were the real and true principles of the physical whole 1.

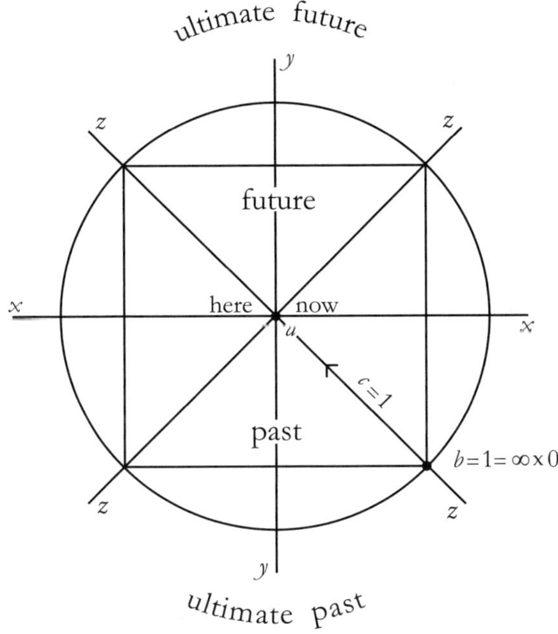

Figure 4.2. Sensible light is the observable part of real, physical light, which travels along the diagonal axis *zz* in the unique forward sense and toward the center *a* at the unique, finite speed $c = 3 \times 10^8$ m/s = 1. The diagonal axis *zz* appears as if it were a simple and undivided (unresolved) line. At the center *a*, sensible light determines a double light cone, the light cone of backward past and the light cone of forward future.

Finally, because the biological selection of speed, wavelength, and frequency of sensible light depends essentially on the genetic constitution of our individual cells (retinal cells), the kind of light that we detect as sensible light is an accident of our peculiar evolution, of our individual destiny, and of our private time generated by our low-energy individual cells.

Metric Description of the Physical Universe

*T*he real, physical universe, defined as the limit of an infinite series of parts, is the greatest magnitude, which we represented one-dimensionally by the limiting-circle b. We define in turn the limiting-circle b as the intersection of the limiting sphere of zero size and the unlimited Euclidean plane of infinite size passing through the limiting sphere's center a. In order to render the complex and indeterminate universe a definite, dimensioned object of our particular perception perceiving backward in time, we assign definite, numerical values to the universe's center a and maximum radius $ab = 1 = \infty \times 0$. These values are selected arbitrarily as a function of the historical accident of our particular senses—of the genetic constitution of our low-energy individual cells. In this way, the real, physical universe is reduced into *our* particular (private), here-and-now, sensible universe of definite, dimensioned radius ab, where a is the inner center and b is the outer limiting-circumference b (see fig. 3.3, p. 61).

On the other hand, because both the center a and the limiting-circumference are necessary sources for describing the physical universe, the maximally distant and different sources $a \neq b$ are ultimately equal and indeterminate: $(a \neq b)(a = b)$. In fact, if we represent respectively the center a and the limiting–circumference b by the concentric circles a and b, and if we correspond every point of the circle a with a unique point of the circle b, we conclude geometrically that they have the same amount of points and hence that they are equivalent. This equality permits us to universalize (by interchanging) the local (particular) properties of a and b, and thus to transform our knowledge of our private, sensible universe seen uniquely from the inner center a into knowledge of the entire physical universe itself, seen both from the inner center a and from the outer limiting-circumference b and hence independent of the accident of our particular senses and their metric definition of the universe.

To the midpoint and center a located at the region of middle scales and designating the here and now (inductive origin) of the space-time diagram let us assign the length 10^{-4} m. We consider 10^{-4} m, which is approximately the size of the smallest organic body—the cell—as the organic unit of length, to which we assign the magnitude 1: 10^{-4} m = 1.[3] With fixed center $a = 10^{-4}$ m and a maximum radius ab, which we take (with respect to extension and division) as $10^{\pm 30}$ times the organic unit 10^{-4} m, we metrically describe the greatest circle or limiting-circle b representing one-dimensionally the physical universe. To the limiting-point b on the limiting-circle b we assign with respect to its horizontal axis xx the cosmological length $10^{-4} \times 10^{30} = 10^{26}$ meters (known as

---

3. The size of the cells in our body varies between 10^{-6} m and 10^{-4} m. For example, the size of our retinal cell is of the order of 10^{-6} m, a length slightly larger than the wavelength of sensible (optical) light, which varies between 10^{-6} m (red light) and 10^{-7} m (blue light). We take, however, as the average size of our cells the length 10^{-4} m, which is also the size of the singled-celled alga, of our lens, and of infrared light emitted from our cells at rest.

Hubble's unit of length and regarded as the radius of the greatest physical body according to extension; we therefore write: $b = 10^{26}$ m.

To the same limiting-point b we assign with respect to its horizontal axis xx the quantum unit of length $10^{-4} \times 10^{-30} = 10^{-34}$ m (regarded as the radius of the greatest physical body according to division—that is, the radius of the smallest physical body), which is the end of the physical universe's extended body when seen from outside as a *space singularity*. We therefore write: $b = 10^{-34}$ m. In total, to the maximally distant, limiting-point b occurring in the region of extreme scales (out there on the greatest circle b at the ultimate past b_p or ultimate future b_f) we assign with respect to its horizontal axis xx the definite numerical magnitudes $b = (10^{26}$ and $10^{-34})$, which measure b's maximum distance from the center a according to extension and division (see fig. 4.3, p. 119).

Let us next take as present time relative to the biological region of scales the time interval 3×10^{-13} s, which is the time required for light to traverse the cell of size 10^{-4} m at the finite speed $c = 3 \times 10^{8}$ m/s. It is also the time for our resting brain to fire one pulse.[4] We consider 3×10^{-13} s as the organic unit of time, to which we assign the magnitude $1: 3 \times 10^{-13}$ s $= 1$. In total, we ascribe to the midpoint and center a the definite, numerical values: $a = 10^{-4}/3 \times 10^{-13} = 1/1$. Assuming that time is proportional to space, we assign to the farthest limiting-point $b = (10^{26}$ and $10^{-34})$ and with respect to its vertical axis yy the respective extreme time intervals $3 \times 10^{-13} \times 10^{+30} = 3 \times 10^{17}$s, and $3 \times 10^{-13} \times 10^{-30} = 3 \times 10^{-43}$ s, such as: $b = (10^{26}/3 \times 10^{17})(10^{-34}/3 \times 10^{-43})$. The limiting-point b designates relative to time the ultimate past (earliest moment or first cause) b_p and the ultimate future (latest moment or final cause) b_f.

---

4. Indeed, our resting individual brain, composed roughly of 3×10^{9} brain cells, fires approximately 3×10^{12} cycles (pulses) per second or 1 cycle (pulse) in 3×10^{-13} of a second. We remark that the frequency of the resting brain's neural action is equivalent to the frequency of infrared light of wavelength $\lambda = 10^{-4}$ m continuously emitted from the resting brain, regarded as a set of brain cells whose average size is of the order of 10^{-4} m.

The greatest time interval 3×10^{17} s (known as *Hubble's unit of time*) is the time required for light emitted from the farthest and earliest (or latest) limiting-point b to travel the maximum distance $ba = 10^{26}$ m at the finite speed $c = 3 \times 10^8$ m/s and to strike the cell at the center a. This time, therefore, expresses the maximum time delay between the emission and the reception of light, which determines the maximum age of the universe and its emitted light when seen as the oldest body.

On the other hand, the smallest time interval 3×10^{-43} s (known as *Planck's unit of time*) is the time required for light emitted from the farthest and earliest (or latest) limiting-point b to travel the minimum distance $ba = 10^{-34}$ m at the finite speed $c = 3 \times 10^8$ m/s and to strike the cell at a. This time, therefore, expresses the minimum delay between the emission and the reception of light, which determines the minimum age of the universe and its emitted light when seen as the youngest body—as the *time singularity*, regarded as the end of the body and its proper age (see fig. 4.3).

Because the finite, numerical magnitudes of the limiting-point $b = (10^{26}/3 \times 10^{17})(10^{-34}/3 \times 10^{-43})$ are maximum magnitudes, and a maximum magnitude is any finite magnitude that is simultaneously (in conformity with the finite–infinite equivalence principle) an infinite magnitude, we conclude that the maximum, finite magnitudes 10^{26} m and 3×10^{17} s are simultaneously infinite magnitudes according to extension measuring an infinite space-time radius, such as: $ab = 10^{26}/3 \times 10^{17} = \infty/\infty$. On the basis of the same principle, we can also conclude that the maximum, finite magnitudes 10^{-34} m and 3×10^{-43} s are simultaneously infinite magnitudes according to division and measure respectively a zero space-time radius, such as: $ab = 10^{-34}/3 \times 10^{-43} = 0/0$. Ultimately, we assign to the universe's unit radius ab and with respect to its horizontal and vertical axes xx and yy the following maximum magnitudes according to extension and division: $ab = 1 = (10^{26}/3 \times 10^{17})(10^{-34}/3 \times 10^{-43}) = \infty \times 0$. These finite magnitudes measure the universe's infinite size and infinite age and also the universe's zero size and zero age. In other words,

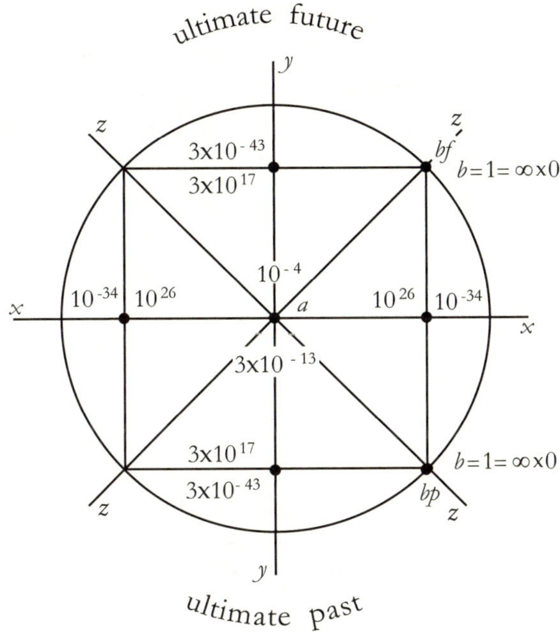

Figure 4.3. In this metric space-time diagram, the physical universe is represented two-dimensionally as the synthetic product (composition) of the outer limiting circumference and the inner unlimited square. The horizontal axis xx represents three-dimensional space measured in meters and the vertical axis yy represents one-dimensional time measured in seconds. Space and time scales are logarithmic, each division representing a change by a factor of 10. We assign to the center a (considered as the here and now of the diagram) the biological length 10^{-4} m taken as the average size of an individual cell in our body and the time interval 3×10^{-13} s, which is the time required for light to traverse the distance 10^{-4} m at the finite speed of light 3×10^8 m/s. We assign to the maximally distant limiting-point b the cosmological length 10^{26} m, regarded as the finite radius of the greatest body, and the quantum length 10^{-34} m, regarded as the finite radius of the smallest body. Assuming that time varies proportionally to space, we assign to the extreme point b occurring in the ultimate past b_p, or ultimate future b_f, the cosmological time 3×10^{17} s, which is the maximum age of the universe thought of as the greatest and oldest body, and the quantum time 3×10^{-43} s, which is the minimum age of the universe thought of as the smallest and youngest body.

these finite magnitudes constitute the finite, dimensioned expressions of the infinite universe according to extension and division and relative to its horizontal space and vertical time. The constant ratio of infinite and zero sizes gives globally a balanced, spherical universe of maximum size equal to the real $1 = \infty/0$, where 1 is both the numeric magnitude that measures the maximum extension of the universe relative to its smallest size (cosmic singularity) and the synthetic unity and proportionality of its infinite and zero magnitudes.

Because the physical universe's complex unit radius $ab = 1 = \infty \times 0$ has simultaneously infinite and zero magnitudes, the limiting-point $b = 1$ has simultaneously infinite and zero distances from the center a. Thus, seen from inside, the limiting-point $b = 1$ is an infinitely distant and inaccessible (transcendent) point releasing light of infinite wavelength and zero frequency; seen from outside and beyond, the limiting-point $b = 1$ is an immediately accessible (immanent) point having a zero distance from the center a and releasing light of zero wavelength and infinite frequency. This means that infinitely distant stars are immediately visible by us if their infinite distance is simultaneously zero and if our cell or brain has the power to perceive both within and beyond the infinite distance of the stars via the universe's cosmic singularity.[5]

Another immanent property of maximum magnitude, besides admitting finite, infinite, and zero magnitudes at the same time, is being both variable and constant. Let us take as the greatest magnitude,

5. Thus, an infinitely distant star is *immediately* visible to us if we perceive the star beyond its infinite distance through its cosmic singularity of zero space and zero time. Such an immediate perception at a distance requires that we have the power to perceive the starlight's infinite and zero wavelengths. Through the star's infinite wavelength, we perceive its infinite distance; through the star's zero wavelength, we perceive its infinite distance *immediately*, and the infinitely distant star as being here and now at a zero space-time distance from us. The finite, dimensioned expression of infinite wavelength is the cosmological length 10^{26} m ($10^{-4} \times 10^{30}$), whereas the finite, dimensioned expression of zero wavelength is the quantum length 10^{-34} m ($10^{-4} \times 10^{-30}$).

according to extension, the cosmological length 10^{26} m. If we multiply 10^{26} by the factor 10^{1}, we obtain the magnitude $10^{26} \times 10^{1} = 10^{27}$, which, although greater than 10^{26}, is equal to 10^{26}:

$$4.4 \qquad (10^{27} > 10^{26})(10^{27} = 10^{26}).$$

If we multiply the greatest, finite magnitude 10^{26} by the factor ∞, we obtain the infinite magnitude $10^{26} \times \infty = \infty$, which is both greater than and equal to the finite magnitude 10^{26}:

$$4.5 \qquad (\infty > 10^{26})(\infty = 10^{26}).$$

The sign ∞ designates the infinite magnitude according to extension, or the maximum limit of an infinite sequence of increasing infinite magnitudes, such as:

$$4.6 \qquad 10^{26},\ldots < \infty, \ldots < \infty + \infty, \ldots < \infty \times \infty, \ldots < \infty^{\infty},\ldots < \infty_{\infty}, \ldots = \infty.$$

Formula 4.5 shows that the infinite variation of the greatest, finite magnitude 10^{26} both modifies and does not modify 10^{26}. Insofar as we extend the finite magnitude 10^{26} without limit, we conclude that 10^{26} is a member of the Euclidean series of parts, which (in conformity with the analytic principle of Archimedes) always has an external magnitude greater than itself. In this sense, the finite magnitude 10^{26} is an indefinitely extensible or variable part (a potential infinite) conditioned by asymmetric time and by the operations of arithmetic. However, insofar as the infinite variation of the finite magnitude 10^{26} leaves this magnitude unchanged, we conclude that 10^{26} is the greatest of all magnitudes, having no magnitude external to itself and greater than itself. In this sense, the finite magnitude 10^{26} is a constant, extensionless, and ageless whole (a cosmic singularity) transcending both the Archimedean principle and the Euclidean sequence of varying parts and hence unconditioned by asymmetric time and the operations of arithmetic.

In order to show that the finite magnitude 10^{26} belonging to the Euclidean series is simultaneously beyond the Euclidean series and hence is the maximum limit of an infinite sequence of increasing magnitudes, we can write formula 4.6 as follows:

4.7 $\quad 10^{26}, \ldots < \infty, \ldots < \infty + \infty, \ldots < \infty \times \infty, \ldots < \infty^{\infty}, \ldots < \infty_{\infty}, \ldots = 10^{26}.$

The same argument is valid for the greatest magnitude according to division—that is, for the smallest magnitude, which is the quantum unit 10^{-34} m. If we multiply 10^{-34} m by the factor 10^{-1}, we obtain the magnitude $10^{-34} \times 10^{-1} = 10^{-35}$, which, although less than 10^{-34}, is equal to 10^{-34}:

4.8 $\qquad\qquad (10^{-35} < 10^{-34})\ (10^{-35} = 10^{-34}).$

If we multiply the smallest, finite magnitude 10^{-34} by the factor $1/\infty$, we obtain the magnitude zero $10^{-34} \times 1/\infty = 0$, which is both less than and equal to the finite magnitude 10^{-34}:

4.9 $\qquad\qquad (0 < 10^{-34})(0 = 10^{-34}).$

The sign 0 designates the infinite magnitude according to division, or the minimum limit of an infinite sequence of decreasing infinite magnitudes, such as:

4.10 $\quad 10^{-34}, \ldots > 1/\infty, \ldots > 1/\infty + \infty, \ldots > 1/\infty \times \infty, \ldots > 1/\infty^{\infty}, \ldots > 1/\infty_{\infty}, \ldots = 0.$

Formula 4.9 shows that the infinite division of the smallest, finite magnitude 10^{-34} both divides and does not divide 10^{-34}. Insofar as we divide the finite magnitude 10^{-34} without limit, we conclude that 10^{-34} is a member of the Euclidean series always having (in conformity with the Archimedean principle) an external magnitude smaller than itself. In this sense, the finite magnitude 10^{-34} is an indefinitely divisible or variable part (a potential infinite) conditioned by asymmetric time and the operations of arithmetic. However, insofar as the infinite variation of the finite magnitude 10^{-34} leaves this magnitude the same, we conclude that 10^{-34} is the smallest of all magnitudes, having no magnitude external to itself and smaller than itself; in this sense, the finite magnitude 10^{-34} is a constant, undivided, and ageless whole (a cosmic singularity) transcending both the Archimedean principle and the Euclidean series of parts conditioned by asymmetric time and the operations of arithmetic.

In order to show that the finite magnitude 10^{-34} belonging to the Euclidean series is simultaneously beyond the Euclidean series and thus is the minimum limit of an infinite sequence of decreasing magnitudes, we can write the formula 4.10 as follows:

$$4.11 \quad 10^{-34}, \ldots > 1/\infty, \ldots > 1/\infty + \infty, \ldots > 1/\infty \times \infty, \ldots > 1/\infty^{\infty}, \ldots > 1/\infty_{\infty}, \ldots = 10^{-34}.$$

On the basis of these considerations, we can define the greatest magnitude according to extension and division as the magnitude that is the contact point b of the inner, open (flat or hyperbolic) part of the universe admitting infinite variation and multiplicity and of the outer, infinitely curved part of the universe (its cosmic singularity) closing the open part, stabilizing the unlimited variation, and unifying the infinitely varying multiplicity of the open part. In short, the greatest and smallest magnitudes together correspond to Plato's complex, geometric figure of *undivided line* (ἄτομος γραμμή), whose undivided and extensionless point (cosmic singularity) assigns unity to the infinitely divided and extended line containing the universe's infinitely varying multiplicity. Ultimately, the greatest and smallest magnitudes satisfy the physical universe's synthetic principle of equivalence, whose varied manifestations are the point–line, whole–part and finite–finite equivalence principles.

Having finished the elementary metric description of the real, physical universe, which we reduced accidentally into a finite, measurable object of our particular perception, we now examine the ontological division of the universe into sensible universe and ideal, real, physical universe. By *sensible universe* we mean the finite, observable part of the physical universe that sensible light reveals to our particular senses; by *ideal, real, physical universe,* we mean the universe existing in itself (τό καθαυτό) as an infinite whole, independent of our particular senses.[6]

6. The ontological division of the universe into sensible universe/physical universe reflects the categorical divisions part/whole, individual/universe, local/global, variable/constant, and center/circumference, whereby the sensible part or individual is located at the midpoint center *a* and the physical whole or universe is located on the limiting-circumference *b,* which is the extremity of the physical universe.

Of the Ontological Division of the Universe into Sensible Universe and Physical Universe

Let us think of our ideal, universal mind, which has the cognitive faculties of universal reason and universal sensibility, as the greatest circle, the limiting-circle b, of center a and radius ab equal to the real $1 = \infty \times 0$. At the low center a, we situate our imperfect, individual mind having the finite faculties of particular sensibility and analytic understanding. The organ of the individual mind is our resting, individual brain (see fig. 4.4). By *resting individual brain* we mean a set of roughly 3×10^9 (3 billion) brain cells embedded in Euclidean space-time of zero curvature and zero force (ground state) and behaving in an analytic and time-conditioned (oriented) manner.[7]

---

7. If we divide the brain's mass, which is roughly 3×10^0 kilograms, by the mass of one brain cell, which we take as being of the order of 10^{-9} kg (one billionth of a kilogram), we obtain the number of cells contained in our brain, which is about 3×10^9 (3 billion) cells. Per contra, if the mass of our body varies between 10^0 kg and 10^2 kg, then the number of cells contained in our body varies proportionally between 10^9 and $100 \times 10^9 = 10^{11}$ (100 billion) cells.

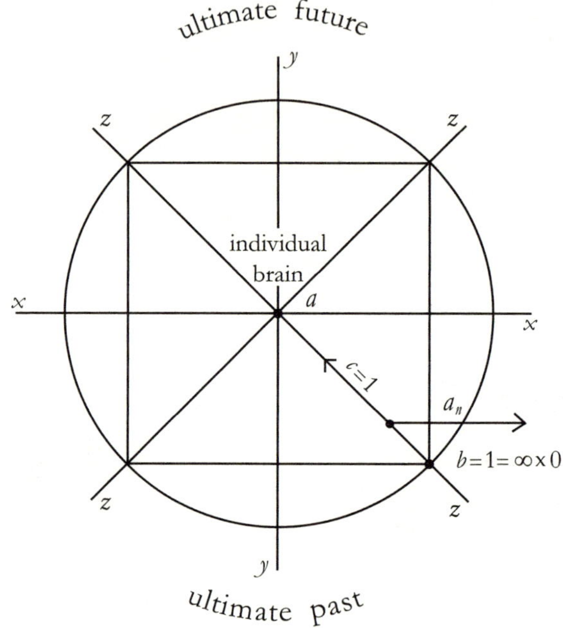

Figure 4.4. The resting individual brain situated at the center and starting point a perceives only a finite portion of the incident, physical light, which we call sensible light, traveling along the simple diagonal axis zz in a unique (forward) sense at the unique, constant, finite speed $c = 1$. Sensible light carries information about only the observable part a_n (the inner and past part) of the physical whole 1. The sensible part a_n we call sensible universe. This part a_n appears to the individual brain at a as accelerating indefinitely away from the center a and toward the limiting-point $b = 1$, where b is the physical whole 1 occurring at the ultimate past (or ultimate future).

Because the quantities of the resting individual brain, such as its size, mass, energy content, temperature, and density, are assumed to possess intermediate magnitudes lying midway between the extreme (maximum) cosmological and quantum magnitudes, we justifiably situate the resting individual brain or brain cell at the universe's region of middle scales at the center a. Thus, we assume that the resting brain cell's average size is of the order of 10^{-4} m (which is also the size of our lens and of the wavelength of infrared light); its mass is of the order of 10^{-9} kg; its density is of the order $10^{-9}/(10^{-4})^3 = 10^0$ gm/cm^3 (that is, 1 gram per cubic centimeter), which is the density of water;[8] its energy content is of the order of 10^{-2} eV (one hundredth of an electronvolt, which is also the energy of infrared light); and its temperature is of the order 10^2 °K (which is also the temperature of infrared light).[9] This low electrical energy exists only so long as the cell carries a slow but unceasing oxidation (burning) of glucose during its resting (ground) state.

Because the low-energy brain cell behaves in its active state in an analytic-categorical all-or-none manner (for example, either it fires or it does not fire a pulse in one–thousandth of a second), we assume that the brain cell is an *individual* either originating or obeying analytic

8. The cell has a mass of the order of 10^{-9} kg if its size (radius) is of the order of 10^{-4} m. Indeed, if we divide the cell's mass 10^{-9} kg by its volume $(10^{-4}$ m$)^3$, we obtain the density of the cell, which is equal to the density of water: $10^{-9}/(10^{-4})^3 = 10^0$ gr/cm^3.

9. In fact, the resting cell, whose average size is of the order of 10^{-4} m and temperature is of the order of 10^2 °K, emits or receives infrared light of wavelength 10^{-4} m, frequency 3×10^{12} cycles/s, and energy 10^{-2} eV.

A cell whose average size is of the order of 10^{-4} m can also be thought of as a lens of the same size whose exchange surface is convex curving infrared light ($\lambda =10^{-4}$ m) toward the cell's center a or concave curving infrared light away from the center a. Curving light toward the center a can be regarded as a method of receiving light, whereas curving light away from the center a can be seen as a method of emitting light.

principles of being. By *analytic principles of being* we mean the principles of self-identity, excluded third, contradiction, inequality, and temporal order.

Let us think of our resting individual brain as an imperfect observer, perceiving only that kind of light, which we called finite, sensible light, out of the total, physical light, the *first light*, coming from all parts of the real, physical universe at all speeds and wavelengths and carrying information about the entire, physical universe, about its inner and outer parts, its past and future. Sensible light of finite speed c and finite wavelength λ carries information about only a finite portion of the real, physical universe—namely, the inner and past part of the real, physical universe, which we call the sensible (observable) universe. The sensible universe appears to our low-energy individual brain at the center a as a time-conditioned Euclidean (or hyperbolic) plane of infinite magnitude having no limiting-point b (no cosmic singularity) to delimit and close it; to unite its discontinuous, consecutive parts; and to transform it into a timeless, real, infinite whole $1 = \infty \times 0$, which is the ideal and real body of the physical universe.

An infinite magnitude without limit is a potential or improper infinite; in reality an improperly infinite magnitude is a finite variable a_n, which, under the action of repulsive gravity R, recedes indefinitely away from the center and starting point $a = 0$ and converges indefinitely toward the end–point $b = 1$ on the limiting-circumference b at an increasing speed (see fig. 4.4, p. 126). We therefore use the variable a_n to designate: (1) the indefinitely accelerating sensible universe with an indefinitely expanding size, which, insofar as it is a fraction of the inacccessible infinite whole, its size and motion can be regarded as zero; and (2) the individual brain generating the indefinitely accelerating and expanding sensible universe. On the other hand, we use the limiting-point $b = 1$ on the limiting-circumference b to designate, depending on the case: (1) the outer, convex part (cosmic singularity) of the limiting-circumference b; and (2) the limiting-circumference b itself (representing one-dimensionally the physical universe) defined as the synthetic unity and contact of the inner, unlimited series a_n (assimilated to the inner, flat or concave part of the limiting-circumference b) with the outer, limiting-point $b = 1$ (assimilated to the outer, convex part of the limiting-circumference b).

When our individual brain is at a state of rest and its energy content is low, it considers the limiting-point $b = 1$ (designating the real, physical universe) as a *transcendent* (absolutely beyond and absolutely inaccessible) thing occurring either before or after the sensible a_n, but never simultaneously and in contact with the sensible a_n. Thus, the real, physical universe numbered by 1 and defined as the simple, transcendent universe of an indefinitely converging sensible universe a_n, is not itself a sensible universe. Similarly, the real, physical universe, defined as the transcendent, *first* and *final cause* of a dynamic series of sensible causes, is not itself a sensible cause. It follows that the transcendent, real, physical universe is a simple (individual) and absolutely supersensible (intelligible) and intellectual point, which we assimilate to the Kantian *noumenon* and to the *transcendent cosmic singularity* of zero extension, which certain relativistic cosmologists call the black hole or the big bang.[10] These last two terms—namely, the Kantian and relativisitic

---

10. Certain relativistic cosmologists attempt to originate in a violent and absurd manner the extended sensible universe from the transcendent, physical universe represented zero dimensionally by an intelligible simple point, and that we call, depending on the case, *noumenon, black hole, big bang,* or *fury God.* But how is it possible to originate extension from something entirely different and contradictory, which is the extensionless point?

Indeed, as we argued in the Fundamental Properties of Cosmic Singularity, note 14 (pp. 70–71), it is impossible to derive the extended sensible line from the extensionless intelligible point, to derive something from nothing, if there is no assumed equality, simultaneity, and immediate contact between them. It follows that there is no transcendent physical universe, no intelligible simple point that violently or absurdly originated the extended sensible universe. In fact, a transcendent physical universe is an absolute nothing, an

terms—are different designations of an absolutely impossible thing, an *absolute nothingness,* occurring nowhere, or uniquely in dimension zero and releasing no light or light having zero speed (or infinite wavelength) and zero frequency. In this case, the transcendent, physical universe assimilated to the transcendent cosmic singularity is not a luminous body but rather a black hole!

absolute irreality or impossibility releasing no light or light of zero speed and zero frequency. It follows that the *real* physical universe is necessarily with respect to zero dimension a *transcendental,* complex point existing simultaneously with the extended, sensible universe, and relatively closing the unlimited sequence of its sensible parts.

Because this transcendental, complex point is both within and beyond the unlimited, sensible universe, it is a *relative nothingness* comprising everything—namely, the property of being a sensible thing belonging to the sensible series and the property of being an intelligible nothing transcending and limiting the unlimited sensible series. Thus, far from being a black hole, this transcendental, complex point (representing zero-dimensionally the real, physical universe) has a maximally luminous mass that releases light at maximum speed and maximum frequency equal to the real $1 = \infty \times 0$.

If we assume that the luminous mass of the real, physical universe is 10^{51} kg, and that the entire mass moving at the finite speed of light $c = 3 \times 10^8$ m/s is converted into energy according to $E = mc^2$, then its total released energy is:

$$E = 10^{51} \times (3 \times 10^{\,8})^2,$$

$$E = 10^{68} \text{ joules.}$$

If the physical universe's maximum age with respect to its vertical axis yy is 3×10^{17} s, then the total energy expenditure of the universe per second is roughly: $10^{68}/3 \times 10^{17} = 3 \times 10^{50}$ joules/s or 3×10^{50} watts.

This maximum energy expenditure defines the maximum instantaneous power or intrinsic luminosity of the universe's greatest mass. Because according to the finite–infinite equivalence principle, which governs the greatest mass, anything finite is simultaneously infinite, the finite power 3×10^{50} W has simultaneously an infinite magnitude: 3×10^{50} W $= \infty$. The infinite magnitude measures the universe's infinite intrinsic luminosity, when seen from outside (from dimension zero) as the greatest luminous mass.

Because there is no sensible, limiting-point $b = 1$, no visible or *sensible cosmic singularity* to visibly unite any two parts of the indefinitely accelerating sensible universe a_n, we conclude that the sensible universe is composed of discontinuous, consecutive, isolated parts (the individuals) deprived of unity, of continuous, immediate motion, and which verify analytic principles of being. Analytic principles therefore are not objective principles of the real, physical whole thought of as prior to and independent of our individual brain, but are instead biological principles of our resting, individual brain expressing the discontinuous and successive manner with which it perceives the continuous and timeless, physical whole. Because the sensible part a_n (the observable portion of the physical whole 1) reflects, as the *effect* of our particular perception, the analytic properties of our time-conditioned individual brain, its content is simple and less than the complex content of the physical whole 1: $a_n < 1$. We then say that the sensible part a_n conditioned by time suffers a loss of content; in fact, what it has lost is the exact measurement and exact perception of the real 1—designating the real, physical universe. We may then regard the positive difference $1 - a_n > 0$ as the error $e = 1 - a_n$ of inexact measurement and perception of the physical whole 1 by the sensible part a_n. The magnitude of this metric and perceptual error can be infinitely small or infinite but never zero (see note 8 in chapter 2, p. 29).[11]

Thus, our low-energy, individual brain or mind suffers from a double error. The first is the illusion of perceiving the timeless, limiting sphere of the permanently rotating (or vibrating), physical universe of radius $ab = 1 = \infty \times 0$ *as if* it were uniquely a time-conditioned, Euclidean (or hyperbolic) sensible plane a_n receding indefinitely away from its

---

11. If the positive difference $1 - a_n$ is simultaneously the magnitude of error e and light's wavelength λ, then the analytic theory of convergent series states that no matter how much the converging a_n contracts $1 - a_n$, it can never reach zero wavelength and hence zero error that permits the exact perception and measurement of 1.

low center and starting point a and indefinitely converging to its high, limiting-point $b = 1$ at an increasing speed. The second error is the onto-epistemological illusion of concluding that this time-conditioned, Euclidean, sensible plane a_n verifying analytic principles of being reveals the exact form and objective nature of the real, physical universe. We can express our individual mind's double error through Plato's famous analogy of an imperfect mind constructing an incomplete copy of the original and taking the incomplete copy *as if* it were the original.

We now come to the question of the speed of recession and expansion of the finite, sensible universe a_n. We assert that v, denoting the speed of recession and expansion of a_n, is proportional to its distance d from the individual brain or mind at the center and starting point a:

4.12 $$v = kd.$$

Here k, the constant of proportionality linking speed to distance, is equal to v/d. This equation is Hubble's law applied to galaxies. In a general manner, this equation is also Hubble's law applied to our sensible universe a_n taken as one galaxy or sequence of galaxies, as one sensible universe or sequence of sensible universes flying away from us at an increasing speed under the force of repulsive gravity R (or under the repulsive action of the original gravitational force F_u). As the sensible universe a_n recedes backward in time and outward in space under the force R, its speed of recession v increases in proportion to its distance d from the center a, regarded by the individual brain as the starting point of rest and minimum speed.[12]

---

12. The relation of proportionality between recessional motion and distance from the starting point a has been known since the time of the ancient Greeks. In his *Physics* VIII (9), $265b_{11}$, Aristotle mentions that motion on a straight line is nonuniform, that things leaving the starting point a (the point of rest, of minimum speed) and converging to the end point b (the point of maximum speed) accelerate the farther away they get from the starting point a and approach the end point b. For Aristotle (who follows the spirit of the age), nonuniform motion on the indefinitely extensible straight line is incomplete

If we consider the distance d as the radius r of the receding, sensible universe a_n, then equation 4.12 becomes:

4.13
$$v = k\,r.$$

This equation states that v, the speed of recession of the sensible universe, varies in proportion to its radius r and that the radius r increases the speed of recession v, which in turn increases the radius r, becoming thereby the speed of expansion of the sensible universe. Thus, v is the speed of recession and of expansion of the sensible universe a_n.

Although repulsive gravity exists at all scales, it becomes visible to our particular senses uniquely at cosmic scales—say, at the scale of 10^6 LY $= 10^{22}$ m.[13] Indeed, a galaxy at one million light-years away and under repulsive gravity is flying from us at 30 km/s $= 3 \times 10^4$ m/s.

Let us next assign these numerical values to the Hubble constant k, which stipulates not only the rate of recession of the galaxy, but also the rate of increase of its rate of recession. In this sense, the constant $k = 30$ km /s/10^6 LY states that a galaxy at one million light-years away increases its speed of recession by 30 kilometers per second every million light-years

and varies with respect to its speed; it is in opposition to uniform motion on the maximally extended circle, which is complete and constant with respect to its speed. Twenty-four centuries after Aristotle, the astronomer Hubble discovered an analogous proportionality between recessional motion and distance from the starting point by observing the acceleration of galaxies relative to us at the starting point and center a.

13. We have the following astronomical measures: 1 LY (one light-year) equals 10^{16} meters and 1Y (one year) equals 3×10^7 seconds; it follows that 10^{10} LY (10 billion light-years) equals 10^{26} meters and 10^{10} Y (10 billion years) equals 3×10^{17} seconds.

distance.[14] Grounded in this empirically established constant applied to distant galaxies beyond one million light-years, we can generalize Hubble's law and apply it to all distances of the logarithmic continuum of distances that start immediately, here and now—that is, from the center $a = 10^{-4}$ m—and terminate at the maximally distant limiting-point $b = 10^{26}$ m $= \infty$. Accordingly, if the distance d from the center a is equal to 10^{-4} m, and if this distance is simultaneously the radius r of the sensible universe a_n, we conclude (on the basis of equations 4.12 and 4.13 and of Hubble's constant $k = 30$ km/s $/10^6$ LY $= 3 \times 10^4$ m/ s/10^{22} m) that the speed of recession and expansion of the sensible universe is:

4.14 $\qquad\qquad v = 3 \times 10^4/10^{22} \times 10^{-4} = 3 \times 10^{-22}$ m/s.

We take this result as equal to zero speed determining the state of rest of the sensible universe having the size of the cell, or the beginning of

---

14. For the universe's speed of recession and expansion, see Jean Heidmann, *Extragalactic Adventure: Our Strange Universe* (Cambridge: Cambridge University Press, 1982), pp. 61–62. According to some scientists at the Carnegie Observatories in Pasadena, the numerical value of the Hubble constant k is 72 km/s/ Mpc. See Charles Seife, *Alpha & Omega* (New York: Penguin Books, 2003), pp. 54–55. Thus, a galaxy whose distance is 1 megaparsec (Mpc)— that is, 3.26×10^6 LY $= 3.26 \times 10^{22}$ m—flies away from us at 72 km/s $= 7.2 \times 10^4$ m/s. If the distance is 10^6 LY $= 10^{22}$ m, then with a Hubble constant $k = 7.2 \times 10^4$ m/s/3.26×10^{22}, the galaxy recedes from us at the speed:

$\qquad v = 7.2 \times 10^4/3.26 \times 10^{22} \times 10^{22} = 2.2 \times 10^4$ m/s $= 22$ km/s,

which is slightly lower than 30 km/s. Both values are equally true because they involve different but equivalent ways for measuring the universe's maximum age. The inverse of the Hubble constant 2.2×10^4 m/s/10^{22} m determines the universe's maximum age t measured *uniquely* with respect to its diagonal axis zz: $t = 10^{22}/2.2 \times 10^4 = 4.5 \times 10^{17} = 1.5 \times 3 \times 10^{17}$ s. On the other hand, the inverse of the Hubble constant 3×10^4 m/s/10^{22} m determines the universe's maximum age t measured with respect to its vertical axis yy: $10^{22}/3 \times 10^4 = 3 \times 10^{17}$ s.

its eccentric acceleration under the force of repulsive gravity.[15] We can use these numerical values to form a new constant $k = 3 \times 10^{-22}$ m/s/10^{-4} m, indicating the rate of minimum recession-expansion and the rate of minimum increase of the rate of minimum recession-expansion of the sensible universe: For example, its speed increases by 3×10^{-22} m per second for every 10^{-4} m.

We consider next the sensible universe of radius 10^{-4} m as the smallest sensible universe with the smallest radius and the smallest speed in the infinite sequence a_n of sensible universes of increasing radii and speeds. This smallest sensible universe we represent by the point a_0 located at the center and starting point a (see fig. 4.5).

Let us imagine that under the eccentric or stretching force of repulsive gravity, this smallest sensible universe a_0 accelerates backward and outward with respect to the center a at a speed that varies in proportion to its distance from a. If the distance d from the center a is

---

15. We already noted (see pp 76–77) that repulsive gravity R originates in the inertial or spatial component of mass: that as a stretching force, it is proportional to the square of distance separating any two mass points. Given the proportionality between space and repulsive gravity, there is a reciprocal generation between them where space generates repulsive gravity and repulsive gravity generates space. Therefore, we do not need an external motive force, a violent big bang, an angry, transcendent God, or an exotic energy to explain and justify the sensible universe's receding motion. In fact, the universe's accelerating recession-expansion is a property of its extended material body, of the inertial or spatial component of its mass.

In order to keep the universe's material body from flying apart, there is an equal opposite force, which is attractive gravity F residing in the gravitational or material component of the universe's body. In total, these contrary forces, or contrary actions of the same complex, indeterminate gravitational force Fu, produce a constant, balanced, force-free body continuously expanding and contracting at the same time, and hence continuously vibrating or rotating. at the maximum speed of light c equal to the real $1 = 0 \times \infty$.

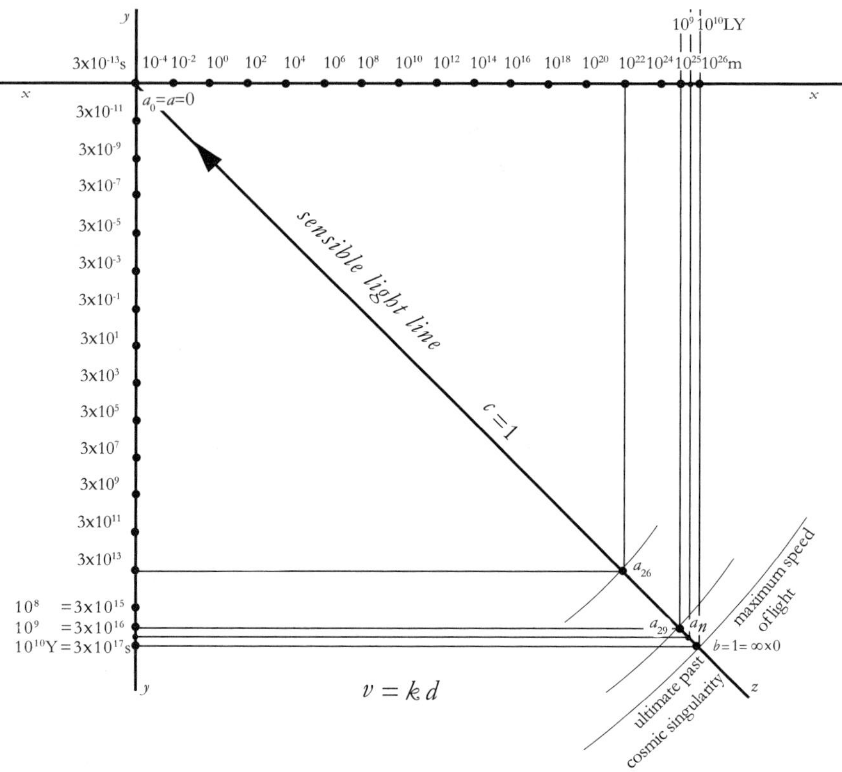

Figure 4.5. With starting point $a = 10^{-4}$ m$/3 \times 10^{-13}$ s, the scales of distance and time expand by 30 powers of 10. Scales of distance are measured in meters, or light-years, on the horizontal axis xx, and scales of time are measured in seconds, or years, on the vertical axis yy. Our individual brain (or brain cell) at the center and starting point a_0 at a is the inductive origin of a sequence a_n of sensible universes of increasing radii and speeds lying on the sensible light line, which is the diagonal space-time ab. Under the unique action of repulsive gravity R, the sequence a_n of sensible universes recedes from $a_0 = a$ and converges toward the end-point $b = 10^{26}$ m$/3 \times 10^{17}$ s on the luminous limiting-circumference b. The starting point $a_0 = a$ denotes the here and now and the end-point b denotes the out there, the ultimate past or ultimate future, the first or final cause of the real, physical universe. Here the figure shows the inner receding part of the physical universe.

10^6 LY (1 million light-years) = 10^{22} m, and if k is 3×10^{-22} m/s/10^{-4} m, then v, the speed of acceleration of the sensible universe, is:

4.15 $\qquad v = 3 \times 10^{-22}/10^{-4} \times 10^{22} = 3 \times 10^{4}$ m/sec or 30 km/s.

We represent the sensible universe accelerating at 30 km/s and located one million light-years away, or having a radius equal to 10^6 LY = 10^{22} m, by the point a_{26}. At this order of magnitude, we can identify the sensible universe a_{26} with a galaxy. A light signal emitted from the point a_{26} and striking our individual brain at the center a needs 10^6 Y (1 million years) = 3×10^{13} seconds to travel the distance 10^6 LY = 10^{22} m at the finite, sensible speed $c = 3 \times 10^8$ m/s (see fig. 4.5).

If the distance d from the center a is 10^9 LY (1 billion light-years) = 10^{25} m on the horizontal axis xx (or 1.5×10^9 LY on the diagonal axis zz), and if k is 3×10^{-22} m/s/10^{-4} m, or 3×10^4 m/s/10^{22} m, then v, the speed of acceleration of the sensible universe, is:

4.16 $\qquad v = 3 \times 10^{-22}/10^{-4} \times 10^{25} = 3 \times 10^7$ m/sec or 30,000 km/s,

$$\text{or}$$

$$v = 3 \times 10^{4}/10^{22} \times 10^{25} = 3 \times 10^7 \text{ m/sec or 30,000 km/s.}$$

We represent the sensible universe accelerating at 30,000 km/s and located one billion light-years away, or having a radius equal to 10^9 LY (or 1.5×10^9 LY), by the point a_{29}. A light signal emitted from the point a_{29} and striking our individual brain at the center a needs 10^9 Y (or 1.5×10^9 Y) to travel the distance 10^9 LY (or 1.5×10^9 LY) at the finite, sensible speed $c = 3 \times 10^8$ m/s (see fig. 4.5).

We consider next this latter sensible universe a_{29} as the least sensible universe a_0 in the infinite sequence a_n of sensible universes, such as: $a_{29} = a_0$ (see fig. 4.6).

If the distance d from the center a is 5×10^9 LY (5 billion light-years) = 5×10^{25} m on the horizontal axis xx (or $1.5 \times 5 \times 10^9$ LY = 7.5×10^9 LY on the diagonal axis zz), and if k has any of the following numerical values:

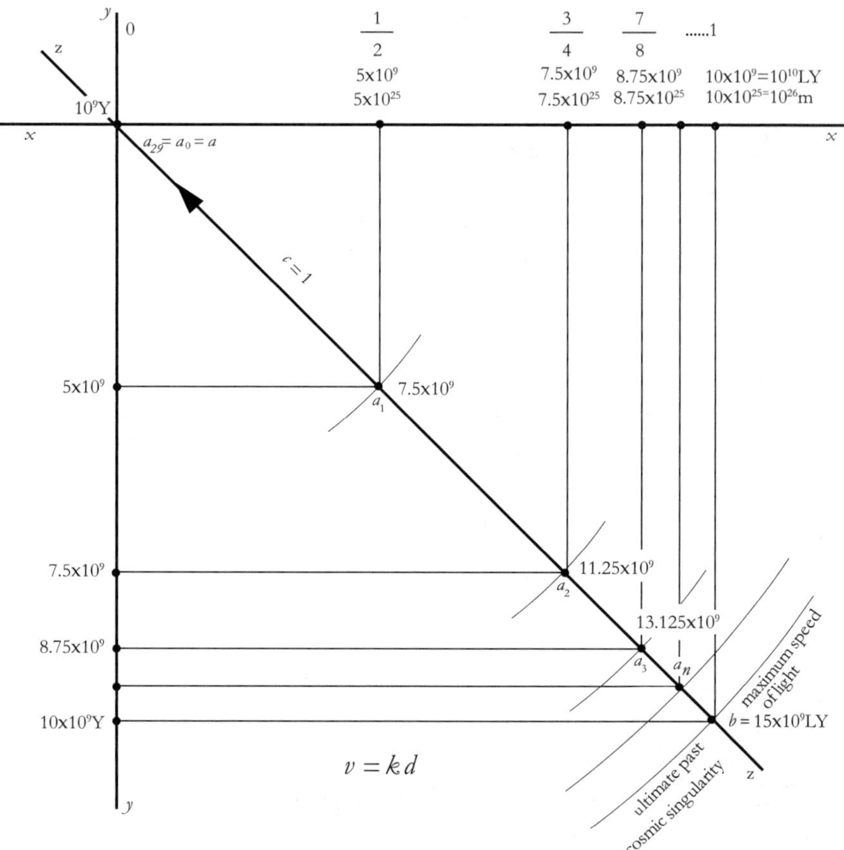

Figure 4.6. In this space-time diagram, the point a_{29} located at 10^9 LY from us is regarded as the least point a_0 in the infinite sequence a_n of points such as: $a_{29}=a_0$. The next point a_1 is located 5×10^9 LY or 7.5×10^9 LY away from us; the point a_2 is located 7.5×10^9 LY or 11.25×10^9 LY away; the point a_3 is located 8.75×10^9 LY or 13.125×10^9 LY away; and the end-point b is located 10×10^9 LY or 15×10^9 LY away.

$$3 \times 10^{-22} \, \text{m/s} / 10^{-4} \, \text{m}; \quad 3 \times 10^4 \, \text{m/s} / 10^{22} \, \text{m}; \quad 3 \times 10^7 \, \text{m/s} / 10^{25} \, \text{m},$$

then v, the speed of acceleration of the sensible universe, is:

4.17 $v = 3 \times 10^{-22} / 10^{-4} \times 5 \times 10^{25} = 15 \times 10^7 \, \text{m/s or } 150{,}000 \, \text{km/s},$

or

$$v = 3 \times 10^4 / 10^{22} \times 5 \times 10^{25} = 15 \times 10^7 \, \text{m/s or } 150{,}000 \, \text{km/s},$$

or

$$v = 3 \times 10^7 / 10^{25} \times 5 \times 10^{25} = 15 \times 10^7 \, \text{m/s or } 150{,}000 \, \text{km/s}.$$

We represent the sensible universe accelerating at 150,000 km/s and located five billion light-years away, or having a radius equal to 5×10^9 LY (or 7.5×10^9 LY), by the point a_1. A light signal emitted from the point a_1 and striking our individual brain at a needs 5×10^9 Y (or 7.5×10^9 Y) to travel forward the distance 5×10^9 LY (or 7.5×10^9 LY) at the finite, sensible speed $c = 3 \times 10^8 \, \text{m/s}$ (see fig. 4.6).

If the distance d from the center a is 7.5×10^9 LY (7.5 billion light-years) $= 7.5 \times 10^{25}$ m on the horizontal axis xx (or $1.5 \times 7.5 \times 10^9$ LY $= 11.25 \times 10^9$ LY on the diagonal axis zz) , and if k has any of the following numerical values:

$$3 \times 10^{-22} \, \text{m/s} / 10^{-4} \, \text{m}; \quad 3 \times 10^4 \, \text{m/s} / 10^{22} \, \text{m}; \quad 3 \times 10^7 \, \text{m/s} / 10^{25} \, \text{m};$$

$$15 \times 10^7 \, \text{m/s} / 5 \times 10^{25} \, \text{m},$$

then v, the speed of acceleration of the sensible universe, is:

4.18 $v = 3 \times 10^{-22} / 10^{-4} \times 7.5 \times 10^{25} = 22.5 \times 10^7 \, \text{m/s} = 225{,}000 \, \text{km/s},$

or

$$v = 3 \times 10^4 / 10^{22} \times 7.5 \times 10^{25} = 22.5 \times 10^7 \, \text{m/s} = 225{,}000 \, \text{km/s},$$

or

$$v = 3 \times 10^7 / 10^{25} \times 7.5 \times 10^{25} = 22.5 \times 10^7 \, \text{m/s} = 225{,}000 \, \text{km/s},$$

or

$$v = 15 \times 10^7 / 5 \times 10^{25} \times 7.5 \times 10^{25} = 22.5 \times 10^7 \, \text{m/s} = 225{,}000 \, \text{km/s}.$$

We represent the sensible universe accelerating at 225,000 km/s and located seven and half billion light-years away, or having a radius equal to 7.5×10^9 LY (or 11.25×10^9 LY), by the point a_2. A light signal emitted from the point a_2 and striking our individual brain at a needs 7.5×10^9 Y (or 11.25×10^9 Y) to travel the distance 7.5×10^9 LY (or 11.25×10^9 LY) at the finite speed $c = 3 \times 10^8$ m/s (see fig. 4.6).

If the distance d from the center a is 8.75×10^9 LY $= 8.75 \times 10^{25}$ m on the horizontal axis xx (or $1.5 \times 8.75 \times 10^9$ LY $= 13.125 \times 10^9$ LY on the diagonal axis zz), and if k has any of the following numerical values:

$$3 \times 10^{-22} \text{ m/s/} 10^{-4} \text{ m}; \ 3 \times 10^4 \text{ m/s/} 10^{22} \text{ m}; \ 3 \times 10^7 \text{ m/s/} 10^{25} \text{ m};$$
$$15 \times 10^7 \text{ m/s/} 5 \times 10^{25} \text{ m}; 22.5 \times 10^7 \text{ m/s/} 7.5 \times 10^{25} \text{ m},$$

then v, the speed of acceleration of the sensible universe, is:

4.19 $v = 3 \times 10^{-22}/10^{-4} \quad \times \ 8.75 \times 10^{25} = 26.25 \ \times 10^7 \text{ m/s} = 262,500 \text{ km/s,}$

or

$v = 3 \times 10^{4}/10^{22} \quad \times \ 8.75 \times 10^{25} = 26.25 \times 10^7 \text{ m/s} = 262,500 \text{ km/s,}$

or

$v = 3 \times 10^7/10^{25} \quad \times \ 8.75 \times 10^{25} = 26.25 \times 10^7 \text{ m/s} = 262,500 \text{ km/s,}$

or

$v = 15 \times 10^7/5 \times 10^{25} \quad \times \ 8.75 \times 10^{25} = 26.25 \times 10^7 \text{m/s} = 262,500 \text{ km/s,}$

or

$v = 22.5 \times 10^7/7.5 \times 10^{25} \times \ 8.75 \times 10^{25} = 26.25 \times 10^7 \text{m/s} = 262,500 \text{ km/s.}$

We represent the sensible universe accelerating at 262,500 km/s and located at 8.75×10^9 LY away, or having a radius equal to 8.75×10^9 LY (or 13.125×10^9 LY), by the point a_3. A light signal emitted from the point a_3 and striking our individual brain at a needs 8.75×10^9 Y (or 13.125×10^9 Y) to travel the distance 8.75×10^9 LY (or 13.125×10^9 LY) at the finite speed $c = 3 \times 10^8$ m/s (see fig. 4.6).

As we recede backward and outward from the (inductive) starting point a under the force of repulsive gravity, we obtain a series a_n of sensible universes lying on the sensible light line and accelerating to the maximally distant end–point $b = 1$ on the physical universe's limiting-circumference b.

Finally, if the distance d from the center a is maximum—that is, 10×10^9 LY (10 billion light-years) $= 10 \times 10^{25}$ m $= 10^{26}$ m on the horizontal axis xx (or $1.5 \times 10 \times 10^9$ LY $= 15 \times 10^9$ LY on the diagonal axis zz), and if k has any of the following numerical values:

$$3 \times 10^{-22} \text{ m/s}/10^{-4} \text{ m}; \ 3 \times 10^4 \text{ m/s}/10^{22} \text{ m}; \ 3 \times 10^7 \text{m/s}/10^{25} \text{ m}; \ 15 \times 10^7 \text{m/s}/5 \times 10^{25} \text{ m};$$
$$22.5 \times 10^7 \text{m/s}/ \ 7.5 \times 10^{25} \text{ m}; \ \text{or} \ 26.25 \times 10^7 \text{ m/s}/ \ 8.75 \times 10^{25} \text{ m},$$

then the speed v of the end–point $b = 1$ is maximum and equal to the finite speed of light $c = 3 \times 10^8$ m/s:

4.20
$$v = 3 \times 10^{-22}/10^{-4} \qquad \times 10 \times 10^{25} = 3 \times 10^8 \text{ m/s} = 300{,}000 \text{ km/s},$$
$$v = 3 \times 10^4/10^{22} \qquad \times 10 \times 10^{25} = 3 \times 10^8 \text{ m/s} = 300{,}000 \text{ km/s},$$
$$v = 3 \times 10^7/10^{25} \qquad \times 10 \times 10^{25} = 3 \times 10^8 \text{ m/s} = 300{,}000 \text{ km/s},$$
$$v = 15 \times 10^7/5 \times 10^{25} \qquad \times 10 \times 10^{25} = 3 \times 10^8 \text{ m/s} = 300{,}000 \text{ km/s},$$
$$v = 22.5 \times 10^7/ \ 7.5 \times 10^{25} \quad \times 10 \times 10^{25} = 3 \times 10^8 \text{ m/s} = 300{,}000 \text{ km/s},$$
$$v = 26.25 \times 10^7/ \ 8.75 \times 10^{25} \times 10 \times 10^{25} = 3 \times 10^8 \text{ m/s} = 300{,}000 \text{ km/s}.$$

We represent the sensible universe accelerating at a maximum speed equal to the finite speed of light $c = 3 \times 10^8$ m/s and located at a maximum distance away by the end–point $b = 1$ on the physical universe's limiting-circumference b. This means that at the maximally distant end-point $b = 1$ there is a sensible universe (or giant galaxy) that has the intrinsic properties of the physical universe's greatest body. Regarded as the greatest body, this sensible universe or giant galaxy $b = 1$ has a maximum radius of the order of $10 \times 10^9 = 10^{10}$ LY (or $15 \times 10^9 = 1.5 \times 10^{10}$ LY) and a maximum luminous mass of the order of 10^{51} kg packed in $(10^{10})^2 = 10^{20}$ solar masses, which make the greatest body a source of maximum radiant power equal to 3×10^{50} watts and hence a star of stars (for the maximum luminosity of the physical universe, see note 10, p. 130). A light signal emitted from the end point $b = 1$ and striking our individual brain at a needs $10 \times 10^9\,Y$ (or $15 \times 10^9\,Y$) to travel the distance $10 \times 10^9\,LY$ (or $15 \times 10^9\,LY$) at the finite speed $c = 3 \times 10^8\,m/s$ along the diagonal axis zz (see fig. 4.6).[16]

16. The finite speed $c = 3 \times 10^8\,m/s$ of sensible light, which is the observable part of light, generates a time delay of $10 \times 10^9\,Y$ (or $15 \times 10^9\,Y$ on the diagonal axis zz) between the emission of light by the end-point $b = 1$ and its reception by the starting point a. This time delay, in turn, determines the age of the emitting source $b = 1$ (the physical universe); it also determines the temperature difference between the physical universe's outer limiting-circumference b and its inner center a and the physical universe's temporal and causal order: $b > a$ or $a < b$, where b designates the outer limiting circumference of the constant physical whole 1 occurring out there at the ultimate past or the ultimate future and thought of as the *first* cause or the *final* end of the accelerating sensible part a_n and a designates the inner center (here and now) where the physical whole 1 appears *as if* it were uniquely the sensible part a_n indefinitely accelerating either away from or toward the center a.

However, as we will show later in this work, the above temporal order and its associated temperature difference between the universe's outer limiting-circumference b and its inner center a are a perceptual illusion of our particular senses, which perceive these differrent parts in a conflicting, consecutive manner. In reality, the real, physical whole 1 is, by means of its unifying cosmic singularity and by virtue of its center–circumference equivalence principle, the same everywhere in space-time, at the inner center a and at the outer limiting-circumference b. Because in a perfectly uniform physical whole 1 there is no time delay, no absolute temporal order and temperature difference, and hence no absolute separation between maximally distant things, between the center a and the limiting-circumference b, the physical whole 1 infinitely accelerates *both* away from and toward the center a, and hence infinitely expands and contracts. The same physical whole 1 infinitely accelerates neither away from nor toward the center a, and hence neither infinitely expands nor infinitely contracts, but instead continuously moves in a circular, uniform motion (rotation) about the center a at the maximum speed of light.

We conclude, then, that the radial acceleration of the sensible part a_n

Because the intrinsic nature of the distance d relative to the starting point a is synthetic and symmetric admitting contrary determinations, it varies symmetrically according to extension and division (inverse extension) under contrary forces R and F (or under the contrary actions of the original gravitaitonal force F_u). It follows that given Hubble's law $v = kd$, or $v \propto d$, the speed v of the receding sensible universe must reflect the symmetric nature of the distance d and therefore must vary both in proportion and in inverse proportion to the distance d. To put it another way, v must increase and decrease by the same factor at the same time:

$$4.21 \qquad (v \propto d)(v \propto 1/d) = v \propto d \times 1/d.$$

Accordingly, as the sensible universe a_n recedes away from the center a, a_n becomes both greater and smaller, faster and slower, despite the fact that at one time we observe and understand, as individual brains, one and only one of the alternatives. At the maximally distant end-point $b = 1$, the rectilinearly receding a_n becomes the permanently rotating physical universe $b = 1$, whose body, by virtue of varying with respect to its size and speed in proportion and in inverse proportion to the distance d, is simultaneously the greatest and the smallest, the fastest and the slowest body having simultaneously with respect to each quantity infinite and zero magnitudes. If heavy is anything moving downward and toward the center a under attractive gravity F and light is anything moving upward and away from the center a under repulsive gravity R, the permanently rotating physical universe, which moves both away and toward the center a at the maximum speed of light $c = 1$, is both the lightest body and the heaviest! On the other hand, the physical universe that continuously rotates about the center a at the maximum speed of light $c = 1$ and at a constant distance from a neither moves away from

indefinitely accelerating *either* away from *or* toward the center a, and hence either indefinitely expanding or indefinitely contracting, is a property of the sensible part a_n of the physical universe and not of the physical universe itself independent of our particular senses.

nor toward the center a, and hence is neither the lightest body nor the heaviest!

Finally, because the physical universe is a source of light permanently rotating at the speed of its emitted light, the cause of light is light itself. This means that light similar to its source of light (the complex universe) is both cause and effect, subject and predicate, verifying thereby the synthetic principle of self-causality. Light is also (as a complex universe) both constant and variable. This means that at one and the same time, the unique and constant speed of light $c = 1$ varies relative to its source's different vertical and horizontal axes (see pp. 108–110), or relative to its source's different motions. For example, insofar as the source of light at the end-point b accelerates (under repulsive gravity R) away from the center a at the speed of light c, the total and real speed of light c striking our individual brain at the center a is zero: $c = c - c = 0$. Insofar as the same source of light accelerates (under attractive gravity F) in the inverse direction—namely, toward the center a at the finite speed of light c,—the total and real speed of light c striking our individual brain at the center a is infinite: $c = c + c = 2c = \frac{1}{0} = \infty$. On the whole, the total and real speed of light c striking our individual brain at the center a has relative to the source's different motions both zero and infinite magnitudes, their product being equal to 1: $c = (c - c)(c + c) = 0 \times \infty = 1$. Our low-energy individual brain at the center a, of course, detects through its particular senses only one magnitude assigned to the speed of light c—namely, that which is equal to 3×10^8 m/s, which we called sensible light or, better, the sensible speed of light, and which we took as the unit of sensible light equal to the simple 1. Generalizing these considerations, we can assert that any unit quantity q of a given kind that appears to our individual brain at the center a as a sensible quantity having at one time the unique and finite magnitude 1 becomes at the end-point $b = 1$ of the physical universe's limiting-circumference b a real, physical quantity—a *first quantity* whose real magnitude 1 is the complex product (or ratio) of infinite and zero magnitudes: $q = 1 = \infty \times 0$.[17]

17. By using Hubble's law $v = kd$, we have shown how at the starting point $a = 10^{-4}$ m the receding body's speed $v = 3 \times 10^{-22}$ m/s, which we take as being equal to zero speed, becomes at the end-point $b = (10^{26}$ m, 10^{-34} m) on the limiting-circle b the maximum speed of light $c = 10^{30} \times 3 \times 10^{-22}$ m/s $= 3 \times 10^{8}$ m/s equal to 1, which we defined as the product (or ratio) of zero and infinite magnitudes: $c = 1 = 0 \times \infty$. We may generalize this particular observation and assert the following for any quantity:

Let q designate a unit quantity of a given kind at rest at the center and start-ing point a: $q = 1$. Let q' designate the same unit quantity in motion receding from the starting point a with a speed v. Let us next assume that q' reflects the symmetric nature of its distance d relative to the starting point a and that q' varies symmetrically according to extension and division and in proportion and inverse proportion to the relativistic factor $\gamma = \sqrt{1 - v^2/c^2}$, such as :

$$(1) \qquad q' = (q \times \gamma)\,(q/\gamma).$$

When at the maximally distant end-point b the speed v of the receding quan-tity q' is equal to the speed of light $c = 3 \times 10^{8}$ m/sec, then v/c is 1 and $\sqrt{1 - v^2/c^2}$ is zero. Because anything q multiplied by zero is zero and anything q divided by zero is infinite, it follows that in conformity with the relation (1), q' has at the same time zero and infinite magnitudes, and thus is maximally decreased and increased:

$$(2) \qquad q' = q \times \gamma \times 1/\gamma \;\rightarrow\; q' = 0 \times \infty = 1.$$

Taking this into consideration, we conclude the following: When the radi-ally receding quantity q' reaches the speed of light $c = 3 \times 10^{8}$ m/sec at the end-point b, it is transformed into a maximally moving and varying quantity having at one time zero and infinite magnitudes: $q' = 1 = 0 \times \infty$. This maxi-mally and symmetrically varied quantity q' whose number is the real 1 we identify with the real, physical quantity. In a general manner, any unit quantity $q = 1$ of a given kind that appears at the center a as a simple, determinate, and undivided quantity (an individual) having the unique magnitude 1, when it reaches the physical universe's limiting-circumference b, where it has the speed of light, becomes what it *really is*—namely, a maximum quantity, a complex, indeterminate, and divided quantity (a universe) whose magnitude 1 is the

We have asserted that the physical universe permanently rotating at the maximum speed of light $c = 1$ is self-contained—that is, both a containing whole 1 transcending the Euclidean series a_n of sensible parts and a contained part a_n belonging to the Euclidean series. Insofar as the physical universe is a containing whole occurring at the end-point $b = 1$ of the limiting-circumference b, it rotates at the speed of light regarded as a maximum speed having no speed greater than itself. We conclude, then, that the maximum speed of light at the end-point $b = 1$ is the end of inequality and temporal order, and hence is the end of radial acceleration involving the existence of temporal order. This means that at the end-point $b = 1$, we have the end of the radially accelerating, sensible universe a_n caused by the stretching force of repulsive gravity and the emergence of a timeless, physical universe equal to 1, which submitted simultaneously to the contrary expanding/contracting actions (forces) of the original force of gravity F_u vibrates or rotates continuously at the maximum speed of light (see fig. 3.5, p. 78). In this sense, we regard the ageless physical universe of maximum size and speed as the *end* of the infinite series a_n of sensible universes of increasing radii, speeds, and constants of cosmic acceleration.

Now, insofar as the physical universe 1 is a contained part occurring within the Euclidean series a_n of sensible parts, it is a sensible universe a_n indefinitely accelerating away from the center a with an increasing radius, speed, and constant of cosmic acceleration. In this case, the physical universe, the greatest and fastest body, has bodies that are greater and faster than itself (greater and faster than the greatest and fastest). It follows that the physical universe as a contained, sensible part neither closes the infinite series a_n of sensible parts nor ends their radial acceleration involving inequality and temporal order. We

product (or ratio) of infinite and zero magnitudes, and which our universal reason conceives as the real quantity: the first quantity reflecting the first body of the real, physical universe.

have, therefore, the following couple of series of constants of cosmic acceleration linking the universe's speed and size:

4.22 $k_0 = 3 \times 10^7$ m/s/10^{25} m; $k_1 = 15 \times 10^7$ m/s/5×10^{25} m;....k_n;...= k_b

$\qquad = 3 \times 10^8$ m/s/10^{26} m = 1

$\qquad k_{29} = 3 \times 10^7$ m/s/10^{25}m; $k_{30} = 3 \times 10^8$ m/s/10^{26} m;... k_n;...= ∞.

The above two series, which are similar to the infinite series of half-distances and to the infinite series of whole numbers, show the following: At one and the same time, the physical universe's greatest constant k_b of cosmic acceleration linking its greatest radius 10^{26} m with its greatest speed 3×10^8 m/s is a limiting-whole k_b transcending the Euclidean series k_n of constants when seen from outside and a contained part k_{30} belonging to the unlimited series k_n of constants when seen from inside. Because the greatest constant k_b has no magnitude external to itself and greater than itself, it refutes the analytic principle of Archimedes. On the other hand, because the finite constant k_{30} has always, in conformity with the analytic principle of Archimedes, an external magnitude greater than itself, it is an indefinitely variable part—a potential infinite— which verifies the analytic principle of inequality and temporal order, in opposition to the greatest constant k_b, which is an infinite whole: an actual infinite governed by the synthetic equivalence principle and zero temporal order (for the properties of the maximum, see also pp. 121–122).

Because our low-energy individual brain at the center a detects (through its particular senses) uniquely the inner, finite part of light (sensible light), we experience the multiplicity $a_n \neq 1$ of the physical universe discontinuously without unity, and hence successively in conformity with the temporal order: $a_n < 1$. No matter how much the expanding sensible part a_n approaches at an accelerating rate its constant, physical whole 1 at the end-point $b = 1$, our individual brain at the center a always perceives the sensible part a_n as being less than

1 and at an infinitely small or infinite distance from 1. In fact, the indefinitely approaching sensible part a_n appears as far off from the physical whole 1 as the least sensible part a_0 of the series of sensible parts located at the center a.

It follows that our low-energy, individual brain perceives the distance separating a_n from 1 *as if* it were a potential infinite, a Euclidean infinite without a cosmic singularity to delimit it and to unite its infinitely separated parts—for example, a_n and 1. It is clear that no accelerating and expanding sensible part a_n *effectively* or *empirically* reaches the real limit 1, where it becomes a permanently rotating physical whole of maximum radius and maximum speed, as long as we are incapable of experiencing the physical whole's cosmic singularity—that is, its supreme unity unifying materially its inner, sensible part a_n with the infinitely distant, physical whole 1. It follows that for our resting, individual brain, the continuous passage of the accelerating sensible universe a_n to the permanent, physical universe 1 rotating at a maximum speed of light $c = 1 = \infty \times 0$ is impossible, apparent, or incomplete.

If our individual brain in its resting state at the center a is incapable of experiencing the physical universe's supreme unity materialized by its cosmic singularity at the limiting-circumference b, then it is sufficient to elevate our brain's resting state of low energy and unity to the level of highest energy and unity at the limiting-circumference b in order to experience the cosmic singularity closing the universe's infinite multiplicity and assigning unity to its infinitely distant and isolated parts. The faculty of perceiving the physical universe's infinite multiplicity synthetically (as an infinite whole) as an infinite unity where *all* is *one* and *one* is *all* we called infinite, synthetic, universal sensibility (see pp. 93–96). This is a faculty that our mind possesses when its organ the brain touches the physical universe's outer cosmic singularity of infinite curvature, infinite energy, and infinite unity. In fact, the infinite unity of the universe's cosmic singularity is manifest within the perceiving brain as infinite, synthetic, universal perception. Thus, the last step for definitively resolving the problem of motion is to find a manner in which to elevate the low-energy content of our resting individual brain to the level of highest energy (the energy of the universe's cosmic singularity), where our brain obtains the power of infinite, universal perception and becomes a universal brain in one-to-one correspondence with its universal mind (νούς).

Different Ways to Elevate the Energy Content of Our Individual Brain

The things in the one world-order are not separated one from the other nor cut off with an ax, neither the hot from the cold nor the cold from the hot.
—Anaxagoras

When our individual brain has in its state of rest a low-energy content, it detects only a finite part of incident light. This finite, sensible light traveling forward along the unique, diagonal axis zz at the unique, finite speed $c = 1$ ensures that we perceive the physical universe in a partial and oriented manner—that is, we perceive the physical universe uniquely from inside as an infinitely regressing Euclidean, sensible universe a_n composed of an infinite multiplicity of consecutive, isolated parts (the individuals) having no cosmic singularity, no global curving force to close its infinite regression, to unite its isolated parts, and therefore to ensure the continuity and immediacy of its communication. *Arithmetic continuum* is what mathematicians call this Euclidean, infinite multiplicity of noncommunicating (isolated) individuals occurring successively in time and having no unity, no intimate bond among them.

In reality, arithmetic continuity is an arithmetic *discontinuity* reflecting the consecutive (contradictory, conflicting) and hence discontinuous manner with which the particular senses of our resting individual brain perceive the multiplicity of the physical world.

Because the energy content of our individual brain at rest depends, for a fixed amount of brain cells, on the size and speed of its brain cells, its energy content and therefore its perceptual power can be maximally increased in two ways: by maximum compression of the size of its brain cells and by maximum acceleration of their speed.

If the electrical energy E and temperature T of a body are inversely proportional to its radius r, then by contracting maximally the cell's body, we increase maximally its energy content and temperature. Let us stipulate the following equations:

4.23 $$E = k \times 1/r \qquad \text{and} \qquad T = k \times 1/r,$$

where k, the constant of energy, is equal to $E \times r$, and k, the constant of temperature, is equal to $T \times r$. As the constant of energy, we take the resting cell's low-energy content, which is of the order of 10^{-2} eV (one hundredth of an electronvolt), and the average radius, which is of the order of 10^{-4} m. As the constant of temperature, we take the resting cell's low temperature, which is of the order of 10^{2} °K (100 kelvins), and the average size, which is of the order of 10^{-4} m (one ten-thousandth of a meter) (see also p. 127). If by some unknown method we contract maximally the cell, and hence we assign to the cell's size the smallest physical length—namely, the quantum length 10^{-34} m—we then increase maximally the cell's energy content and temperature:

4.24 $$E = 10^{-2} \times 10^{-4} \times 1/10^{-34} = 10^{28} \text{ eV or } 10^{19} \text{ GeV (gigaelectronvolts)}[18]$$

18. GeV (gigaelectronvolts) means 10^{9} (1 billion) electronvolts. We may similarly vary the constant k and take as a resting cell the retinal cell of radius 10^{-6} m and energy content 10^{0} eV (which is simultaneously the quantic energy of optical red light of wavelength $\lambda = 10^{-6}$ m and temperature 10^{4} °K). In this case, if the radius of the retinal cell is contracted into the quantum size of 10^{-34} m, its energy content is:

$$E = 10^{0} \times 10^{-6} \times 1/10^{-34} = 10^{28} \text{ eV.}$$

4.25 $$T = 10^2 \times 10^{-4} \times 1/10^{-34} = 10^{32}\ {}^{\circ}\text{K}.$$

The above highest energy of highest temperature (which physicists call *Planck energy*) is the energy of maximum unification realizing the highest unity—that is, the maximum synthesis of maximally different and distant things. If we consider the smallest, finite length 10^{-34} m as equal to zero length (see p. 118), then the finite energy 10^{28} eV of finite temperature $10^{32}\,{}^{\circ}$K is equal to infinite energy of infinite temperature, which is the energy of the physical universe's outer cosmic singularity of infinite curvature: $10^{28} = \infty$ and $10^{32} = \infty$.

We know no method for contracting the cell to the smallest quantum length, which is the finite, dimensioned size of the extensionless cosmic singularity. The ancient Indian practice of Yoga, however, aims at reaching this limiting point of zero extension occurring in zero dimension through mental concentration. By the yogic method, the mind contracts the extension of its organic body (brain, brain cell), containing all multiplicity and all activity, into one single extensionless and dimensionless rest point. Once this maximum mental concentration and organic contraction in a rest point is achieved, the mind is considered to have touched the physical universe's transcendental cosmic singularity of infinite curvature, infinite energy, and infinite unity, and therefore to have become an infinite, universal brain or mind possessing the faculty of infinite, synthetic, universal sensibility. The compression of the extension containing the fluctuations of the mind's organic body must be relative (partial) and not absolute (complete). As we discussed in chapter 3 concerning the fundamental properties of cosmic singularity, the logical problem posed by the realization of maximum unity is how to maximally compress the extension of the body and thus ensure the continuity and immediacy of communication among its isolated parts without destroying the extended body itself. To put it in another way, how can we obtain a maximally concentrated and dense body without destroying its organic properties relating to the density of water?

The second way of maximally increasing the energy content of our individual brain is by accelerating the speed of its constituent cells to the maximum speed of light. If the energy content of the cell's body varies in proportion to its speed, then for a given unit mass it is sufficient to maximally increase the cell's speed in order to maximally increase its energy content. Let us take the equation of special relativity theory:

4.26 $$E = mc^2.$$

If we accelerate the cell's rest mass of the order of 10^{-9} kg (one billionth of a kilogram) to the finite speed of light c, then the energy stored in its rest mass is:

4.27 $E = 10^{-9} \times (3 \times 10^8)^2 = 10^8$ J (joules of energy), or about 10^{27} eV (electronvolts),

which is near the highest energy (the energy of supreme unity) of all-is-one and one-is- all.[19] This highest energy, in addition to being the energy of supreme unity, is the energy of infinite light allowing universal perception, as well as the energy of permanent life and continuous motion.[20] Indeed, this energy allows the cell with a given instantaneous

---

19. The conversion of joules of energy to electronvolts is done on the basis of 10^0 joule = 10^{19} electronvolts.

20. By *infinite light* with respect to speed we mean the infinite part of physical light that travels along the horizontal space axis xx at infinite speed and along the vertical time axis yy at zero speed (see pp. 108–110). We can similarly define *infinite light* as the hypersensible part of light that travels beyond the space and time axes xx and yy, and hence through the cosmic singularity at infinite and zero speeds. Generally speaking, anything that moves uniquely through space at infinite speed is the fastest body that travels an infinite distance in one unit of time or one unit of distance in zero time. If we take as infinite distance the greatest cosmological length 10^{26} m = ∞ and as finite time the organic unit time 3×10^{-13} s = 1 (see p. 117), then the finite, dimensioned magnitude of infinite speed is: $10^{26}/3 \times 10^{-13} = 3 \times 10^{38}$ m/s = ∞.

Anything that moves uniquely through time at zero speed is the slowest body, equivalent to the body at rest, which travels one unit of distance in infinite time or a zero distance in one unit of time. If we take as finite distance

power supply to operate for a maximum or infinite time. For example, if an active brain cell operates with an instantaneous power supply of about 3×10^{-10} W or 3×10^{-10} J/s , then in order to be active continuously and for a maximum time—say, 10^{10} years or 3×10^{17} s—the cell's total

the biological unit length 10^{-4} m = 1 and as infinite time the greatest time interval 3×10^{17} s = ∞, then the finite, dimensioned magnitude of zero speed is: $10^{-4}/3 \times 10^{17} = 3 \times 10^{-22}$ m/s = 0.

By *infinite light* with respect to wavelength we mean the infinite part of physical light that travels along the horizontal space axis *xx* with zero wavelength and along the vertical time axis *yy* with infinite wavelength. We can equally define infinite light as the hypersensible part of light that travels beyond the space and time axes *xx* and *yy*, and hence through the cosmic singularity with zero and infinite wavelengths. Light has zero wavelength, which we can call *hyper-γ* light, when its wavelength λ has the smallest size equal to the quantum length (space singularity) 10^{-34} m. We write therefore: $\lambda = 10^{-34}$ m = 0. On the other hand, light has infinite wavelength, which we call *hyper-radio* light, when its wavelength λ has the greatest size equal to the cosmological length 10^{26} m. We then write: $\lambda = 10^{26}$ m = ∞. The ratio of light's sensible speed $c = 3 \times 10^8$ m/s = 1 to its shortest wavelength $\lambda = 10^{-34}$ m = 0 determines light's infinite frequency, whose finite, dimensioned number is $3 \times 10^8/10^{-34} = 3 \times 10^{42}$ cycles per second or 1 cycle in zero time, that is, in 3×10^{-43} s. We write then: $f = 3 \times 10^{42}$ cycles/s = ∞. This constitutes the sensible definition of infinite frequency relative to light's sensible speed $c = 3 \times 10^8$ m/s. The ratio of light's sensible speed $c = 3 \times 10^8$ m/s = 1 to its longest wavelength, which we can call hyper-radio light $\lambda = 10^{26}$ m = ∞, determines light's zero frequency, whose finite, dimensioned number is $3 \times 10^8/10^{26} = 3 \times 10^{-18}$ of a cycle per second or 1 cycle in maximum time—that is, in 3×10^{17} s. We have thus: $f = 3 \times 10^{-18}$ cycle/s = 0. This constitutes the sensible definition of zero frequency relative to light's sensible speed $c = 3 \times 10^8$ m/s.

Infinite light having simultaneously infinite and zero frequencies gives a spherical or universal perception of the object (universe). Indeed, the maximally energetic brain cells, by virtue of emitting or intercepting hyper-radio

power supply must be $3 \times 10^{-10} \times 3 \times 10^{17} = 10 \times 10^{7} = 10^{8}\,$J or 10^{27} eV;[21] this amount of energy is already stored in the cell's rest mass (see formula 4.27). With a maximum amount of electrical energy stored in the cell's rest mass, the cell has a maximum or infinite time (in the past and in the future) or, to put it another way, is timeless.[22]

Formulae 4.23 and 4.26 show that we can increase the brain cell's energy content and attain the transcendental, cosmic singularity of

light ($\lambda = 10^{26}\,$m $= \infty$) of zero frequency, perceive the infinite multiplicity of things in infinite space-time, whereas by virtue of emitting or intercepting hyper-γ light ($\lambda = 10^{-34}\,$m $= 0$) of infinite frequency, they perceive the infinite multiplicity of things in zero space-time—that is, through the cosmic singularity and hence as unity, as one thing: as real, infinite whole 1.

Thus, infinite light involving the conjunction of zero frequency with infinite frequency enables the high-energy brain cells to perceive the unity of the infinitely many in infinite space-time and therefore to have a universal and simultaneous perception of the infinitely extended and infinitely divided physical universe.

Because the maximally energetic brain cell emits and intercepts infinite light of infinite frequency, its computational power and storage (memory) capacity are infinite. For example, if the high-energy brain cell emits 3×10^{42} cycles (vibrations) encoding 10^{42} parts (bits) per second, we conclude that it has a maximum or infinite computational power. If the high-energy brain cell is continuously active during a maximum or infinite time of the order of $3 \times 10^{17}\,$s, then the maximum or infinite storage (memory) capacity of the cell is: $3 \times 10^{42} \times 3 \times 10^{17} = 10 \times 10^{59} = 10^{60}$ cycles encoding 10^{60} parts. If each cycle encodes two parts (for example, the beginning a and the end b), then the totality of parts encoded in 10^{60} cycles is 2^{n} parts, where $n = 10^{60}$ parts. Because this dimensionless number is the finite expression of the cell's infinite storage or memory power, we write: $n = 10^{60} = \infty$.

21. The power required for an active nerve cell to fire 10^{3} pulses per second, or 1 pulse in $10^{-3}\,$s (one thousandth of a second), is roughly 3×10^{-10} of a joule/s or 3×10^{-10} of a watt.

22. To be timeless means to have both infinite age—that is, $3 \times 10^{17}\,$s $= \infty$— and zero age (time singularity), which is $3 \times 10^{-43}\,$s $= 0$.

highest energy and highest unity either by contracting the cell to the smallest size (the quantum length 10^{-34} m) or by accelerating the cell to the highest speed, the speed of light c. Because both ways are equivalent and mutually dependent, the realization of one way necessitates the other. Thus, if a body (say, a brain cell) has a radius r when it is at rest, then when it is moving with a speed v equal to the speed of light c, its effective radius r' contracts into zero in conformity with the relativistic relation $r' = r\sqrt{1 - v^2/c^2}$, where r is the radius of the cell at rest and r' is the radius in motion.[23] But if zero radius is the radius of the physical universe's transcendental cosmic singularity, it follows that when the brain cell accelerates to the speed of light, it touches, similar to an Indian Yogi in maximum mental concentration, the physical universe's extensionless and dimensionless cosmic singularity of highest energy, unity, and universality.

23. Because on the whole all variation is symmetric, in reality r' varies symmetrically in proportion and in inverse proportion to the relativistic factor γ:

$$r' = r \times \gamma \times 1/\gamma = 0 \times \infty = 1.$$

This shows that when the cell accelerates to the speed of light c, its radius r' is maximally both contracted and extended so that it becomes an ideal, *first* quantity, a maximum r', defined as the product of zero and infinite magnitudes and whose number is the real $1 = 0 \times \infty$ (see note 17, p. 145).

When the cell's size is maximally contracted into an extensionless point (cosmic singularity), the cell emits or intercepts hyper-γ light ($\lambda = 10^{-34}$ m) of maximum frequency of the order of 3×10^{42} cycles/s. When the cell's size is maximally expanded into the infinitely extended line, the cell emits or intercepts hyper-radio light ($\lambda = 10^{26}$ m) of minimum frequency of the order of 3×10^{-18} cycle/s. The synthesis of both kinds of light enables the maximally extended point cell to have a universal action on the object and a universal perception of the object, to travel and perceive the object's infinite space-time at once in zero space-time (see also note 20, pp. 153–155).

We call the *cosmic* or *universal* brain the individual brain composed of individual cells at rest elevated to the level of highest energy, unity, and universality, which is the physical universe's cosmic singularity residing on the outer limiting-circumference *b*. But is such an elevation from the inner center *a* (the place of the individual brain) to the outer limiting-circumference *b* (the place of the universal brain) possible? To put it another way, is it possible for the individual brain to concentrate, by maximum compression or acceleration of its cells, an infinite amount of energy within its finite body without disintegrating into chaotic nothingness? Is it possible for our individual brain to experience the supreme unifying cosmic singularity of the physical universe without being burned by the universe's singularity of infinite energy and infinite temperature?

This experience is possible because the physical universe itself accomplishes such a synthetic, infinite act! Indeed, it is a fundamental property of the balanced and symmetric physical universe to accumulate an infinite amount of energy through its maximally unifying cosmic singularity without being burned by its infinite temperature. In fact, the zero energy of the cold infinitely extended physical universe exactly counteracts the infinite energy of the physical universe's hot singularity. In total, the cold, infinitely extended universe of zero energy and the hot extensionless singularity of infinite energy are maximally different and mutually neutralized parts of the *same* greatest body defined with respect to the unit quantities of energy and temperature, as the constant and balanced ratio of infinite and zero magnitudes. Thus, far from disintegrating the cold, material body of the infinitely extended physical universe, the hot, infinitely compressed and dense singularity ensures the immediate communication of its infinitely distant, isolated parts; the uniform and just distribution of its matter-energy-information; the cause and permanence of its life and motion; the universality of its laws; and the limit, substance, and certainty of its indefinitely varying parts. As a contained member of the physical universe, which is simultaneously equivalent to the physical universe, our brain participates in this fundamental synthetic property of the physical universe, and as a greatest body, our brain accumulates an infinite amount of energy within its finite body without burning or crushing its body! Such an extraordinary synthesis happens as discussed below:

Let us think of our brain as a greatest circle b of center a and unit radius ab. Relative to the inner, individual observer O at the center a, the brain is a low-energy *individual brain* that occurs at the inner center a (the cold region of zero curvature, zero energy, and zero force) and has the state of rest or of indefinite acceleration denoted by the finite variable a_n (see fig. 4.7). Any unit quantity q of a given kind belonging to the individual brain's unit cell (for example, the size of the cell) has a unique magnitude, which we take as equal to the finite 1: $q = 1$.

Because the low-energy individual brain detects only a finite part of incident light, which we call sensible (optical) light of finite wavelength $\lambda = 10^{-6}\text{m} - 10^{-7}\text{m}$ traveling in a unique sense (that is, forward) along the diagonal axis zz at the unique finite speed $c = 3 \times 10^8 \text{ m/s} = 1$, it perceives the physical universe 1 *uniquely* from inside, and hence *as if* it were a time-conditioned, Euclidean part a_n indefinitely converging to the maximally distant end-point $b = 1$ on the greatest circle b. The end-point $b = 1$ represents the real, physical universe with a cosmic singularity rotating at the maximum speed of light $c = 1 = 0 \times \infty$ and regarded both as a knowing subject—a universal brain—and as a known object—a physical universe. Thus, the low-energy individual brain perceives the Euclidean multiplicity $a_n \neq 1$ of the physical world discontinuously and successively such as $a_n < 1$ in conformity with the analytic principle of inequality and temporal order—and thus without a cosmic singularity to unite its parts and ensure their continuous, immediate communication.

We conclude, then, that our individual brain's perception of the physical universe at the center a is time-conditioned—that is, it is oriented, partial, and asymmetric, verifying analytic principles of being.

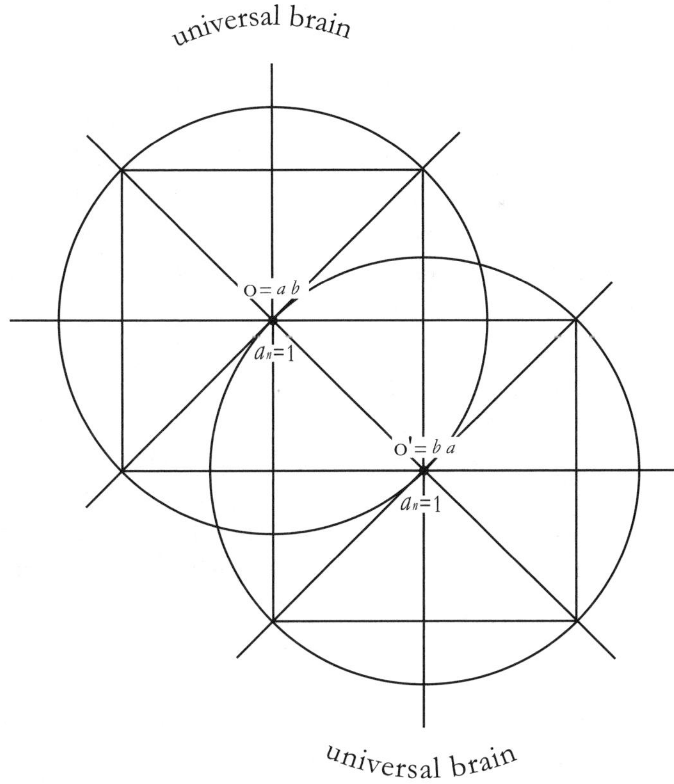

Figure 4.7. In this nonmetric space-time diagram, the inner, individual observer O at the low center *a* experiences our brain as a finite, individual brain at rest or at indefinite acceleration, denoted by a_n. At the same time, the outer, universal observer O′ on the high limiting-circumference *b* experiences our brain as an infinite, universal brain denoted by the real 1 and moving at the maximum speed of light $c = 1 = 0 \times \infty$ on the limiting-circumference *b*. On the whole, our brain is at any point in space-time a complex infinite whole, which is both an individual a_n at *a* and a universal 1 on *b* verifying the synthetic equality $a_n = 1$. This equality permits the indefinitely accelerating individual brain a_n to reach 1, the state of universal brain and experience 1—the sum total of the physical universe's parts.

Now, relative to the outer, universal observer O' on the outer limiting-circumference b, the same brain is a high-energy *universal brain* denoted by the real 1 and rotating at the maximum speed of light $c = 1 = 0 \times \infty$ on the outer limiting-circumference b (the hot region of cosmic singularity of infinite curvature, infinite energy, and infinite force). Here, any unit quantity q of a given kind belonging to the universal brain's unit cell (for example, the size of the cell) has an extreme or maximum magnitude equal to the real 1 and defined as the product of infinite and zero magnitudes: $q = 1 = \infty \times 0$. We have called it ideal or first quantity, analogous to the physical universe's first body (see note 17, p. 145). Because the high-energy universal brain emits or intercepts the entire physical light, the *first* light, emitted from the end-point $b = 1$ and traveling through and beyond space-time along all axes, in all directions, at all speeds and frequencies, it perceives the real, physical universe 1 universally (spherically) from all senses—its inside and its outside, its past and its future—and thus as it *is* in itself (τό καθαυτό): that is, as a balanced, infinitely extended point, an infinite whole that vibrates and rotates continuously under contrary forces at the maximum speed of light $c = 1 = 0 \times \infty$.[24] The real, infinite whole 1 is governed by the synthetic principle of equivalence and zero temporal order, which is materially realized by the infinite whole's cosmic singularity. Thus, the high-energy universal brain, by virtue of touching the world's unifying singularity, perceives its Euclidean multiplicity $a_n \neq 1$ continuously and simultaneously, such as $a_n = 1$—that is, with a cosmic singularity to unite its parts and to ensure their continuous, immediate communication. As a matter of fact, it would have been impossible for the expanding

---

24. For example, the high-energy brain cell (retinal cell) contracted maximally to the quantum size 10^{-34} m (space singularity) has the power to emit and intercept the entire physical light (*first light*) flowing along all axes, in all directions, at all frequencies (that is, at zero and infinite frequencies), and carrying information about the entire physical universe, its inner (infinitely extended) and outer (extensionless) parts, its past and its future.

sensible universe a_n, which has traveled the distance a_n, to *continue* traveling and expanding to 1 if there were no cosmic singularity to unify a_n with 1.

Considering the maximally different observations O and O′ as equally real and true, we conclude that our brain is globally a complex, infinite whole—a finite body with an infinite energy and infinite force occurring at any time within the infinitely regressing Euclidean space-time a_n at the inner center a and beyond the infinitely regressing Euclidean space-time on the outer limiting-circumference b of zero space-time (see fig. 3.4, p. 65 and fig. 4.7, p. 159).

Because our universal brain (the greatest sphere) occurs at any time both at the low center a and on the high, limiting-circumference b, it follows that at a particular time the inductive elevation of our individual brain from the low center a to the high limiting-circumference b is comprehensible and necessary. In fact, the inductive progression of our resting or accelerating individual brain a_n at a to b—the place of the universal brain 1 rotating at the maximum speed of light $c = 1 = 0 \times \infty$—is geometrically and logically necessary if, and only if, the center a and the limiting-circumference b are equivalent in the perfectly uniform, infinite sphere of our physical universe and universal brain (see pp. 63–64). The equality between a and b permits the replacement of a by b, and hence the passage from a to b, the extension of a to b, or the transformation of a into b.[25] If we resolve (decompose, divide) the

---

25. If we assign to the center a the property a_n (the property of being an indefinitely varying part, or an unlimited series of parts) and to the limiting-circumference b the property 1 (the property of being the constant, real 1—that is, the limit and total sum of an unlimited series of parts: in other words, an infinite whole—a physical universe having infinite and zero magnitudes), then the center–circumference equivalence principle becomes our known finite–infinite equivalence principle, which grounds continuous motion from the *unlimited* series a_n to the *limit* 1 within the perfectly uniform, infinite sphere of the physical universe. The primitive spatial contrariety center a–circumference b determines the physical universe's vibratory and rotational motion emerging

equivalence $a = b$ into contrary (opposite and coexisting) inequalities, temporal orders, or rectilinear accelerations, we then have the following formula:

4.28
$$(a = b) = (a < b)(b < a),$$

where a designates the low center (the place of the individual, of the incomplete part a_n) and b designates the high limiting-circumference b (the place of the physical universe, of the complete whole or maximum 1).

Formula 4.28 defines the equality of a and b as the synthetic product of contrary inequalities and temporal orders regarded as contrary parts of an impartial, synthetic whole free of the partiality of the part. The inequality $a < b$ designates the progressive temporal order generating, under repulsive gravity R, the outward acceleration (accelerating expansion-progression) of the universe from a to b. The contrary inequality $b < a$ designates the regressive temporal order generating, under attractive gravity F, the inward acceleration (accelerating contraction-regression) of the universe from b to a. If we arbitrarily regard one of these two temporal orders or rectilinear accelerations as generating a loss of content—for example, of either a or b—then necessarily the converse temporal order restores the lost part.

Finally, we can define the equality of a and b as the negation of contrary inequalities and temporal orders leading to the negation of contrary rectilinear accelerations, of accelerating expansion and accelerating contraction, of accelerating progression and accelerating regression: Mutually negated, they generate a neutral, forceless, and timeless physical universe that is free of rectilinear acceleration and which is neither expanding nor contracting, neither progressing nor regressing:

---

as the balanced ratio of contrary, radial accelerations away from and toward the fixed center a and therefore about the fixed center a, under the contrary expanding and contracting, or progressing and regressing, actions of original gravity: $F_u = (F = R)$.

4.29
$$(a < b)(b < a) \rightarrow (a < b)'(b < a)'.$$

The left side of the formula defines the physical universe of maximum unit radius *ab* as the synthetic product of contrary temporal orders and rectilinear accelerations, such that the complex universe is both expanding and contracting, progressing and regressing. The right side of the formula defines the physical universe as the product of contrary temporal orders and rectilinear accelerations, which are both negated such that the impartial and balanced universe is at rest, neither expanding nor contracting, neither progressing nor regressing. Ultimately, as we have concluded before, the real, physical universe is in permanent rotation and vibration at the maximum speed of light. This is the only kind of motion that does not involve a change of position, and therefore it reconciles motion with rest.

Now, what our low-energy individual brain a_n perceives at one particular place and time at its center *a* and through the finite, sensible part of light is one and only one arbitrarily selected temporal order out of the totality of temporal orders. Accordingly, our individual brain a_n perceives the accelerating expansion of the sensible universe a_n at *a* in conformity with the temporal order:

4.30
$$a < b.$$

We qualify this temporal order as progressive and inductive because the individual brain a_n at the center *a* such as $a_n = a$ is chronologically prior to the universal brain numbered by the real 1 on the limiting-circumference *b*. Indeed, the *end* of inductive progress is the real 1 on *b* and hence the maximum extension of the sensible part a_n experienced by the individual brain a_n at *a* into the real, physical universe 1 experienced by the universal brain 1 on *b*.

Our individual brain also considers this progressive temporal order *as if* it were the immanent and real property of the physical universe. However, because $a < b$ is only an incomplete, sensible part out of the complete totality of parts $(a = b) = (a < b)(b < a)$ that determines the perfectly uniform, infinite sphere of the timeless, physical universe, this incomplete sensible part cannot be an immanent and real property of the timeless, physical whole 1; it is instead a property of the sensible part a_n at the center a accelerating to the real, physical whole 1 on the limiting-circumference b. In fact, outward acceleration (accelerating expansion-progression) is an accidental and apparent effect of the partial, discontinuous, and successive manner with which our low-energy individual brain perceives (through the finite, sensible part of light) the physical universe's totality of temporal orders and radial accelerations.[26]

It follows that the immanent and real property of the permanent physical whole 1 is that of being the constant ratio and synthetic unity of contrary temporal orders and radial accelerations:

4.31 $$1 = (a < b) / (b < a), \qquad (a < b) = (b < a).$$

Attaining this synthetic unity occurring on the limiting-circumference b (the place of the physical universe with a cosmic singularity being present at all points of the physical universe at all times) is also the *first cause, founding principle,* and *final end* of the incomplete, sensible part a_n accelerating to the limiting-circumference $b = 1$ for its perfection.

26. Finite, sensible (optical) light having the unique, finite wavelength $\lambda = 10^{-7}$ m and propagated at the unique, finite speed $c = 3 \times 10^8$ m/s at the unique, finite frequency $f = 3 \times 10^{15}$ cycles/s introduces a time delay between distant things, between receptor and emitter of light, so that we experience the symmetric distance and difference between things as being a contradiction and a succession. An ideal, universal observer (universal brain) who emits or intercepts the entire physical light, the *first light* propagated at both zero and infinite frequencies and hence at the maximum frequency $f = 0 \times \infty = 1$, perceives the infinitely many and distant things immediately through the universe's cosmic singularity and hence free of time delay, free of contradiction and succession.

• 5 •

The Meaning of Acceleration

The Meaning of Acceleration

Anything which is coming into being is incomplete
and in progress toward its principle.
—Aristotle, Physics

Acceleration in a straight line requiring the existence of a temporal order is not an essential and real property of the physical universe; rather, it is an accidental or apparent property generated by our individual brain at rest and projected outward into the physical universe. In fact, our low-energy individual brain at the center a perceives through the sensible part of light the physical universe on the limiting-circle b uniquely from inside *as if* it were a time-conditioned sensible part a_n accelerating toward b, the place of its perfection, where it emerges as a physical universe numbered by the real 1. We can then consider the real 1 governed by the synthetic principle of equivalence as the *first* cause, the *final* end, and the *first* founding principle of the accelerating sensible part a_n—that is, of our accelerating galaxies and individual brains, of our accelerating technology and empirical science.

Different Meanings of the Real 1

Οὔτε δε ἡ μονάς ἀριθμός, ἀλλά ἀρχή ἀριθμοῦ.
—Theon of Smyrn (Neoplatonist philosopher)

*T*he real 1 is both a number and a principle of number. As a number (unit), the real 1 designates the physical body as *one* thing. As a principle of number, the real 1 designates the physical body as the *unity* of an infinite multiplicity of things. In this sense, the real 1 constitutes the synthetic equivalence principle governing the physical body and its corresponding number. Similarly, as a number, the real 1 designates the greatest quantity of one kind according to extension and division. As the synthetic equivalence principle, the real 1 is the synthetic unity (proportion) or constant ratio of different parts (quantities or magnitudes) ensuring the constancy of the physical body. The real 1 has a physical meaning if it denotes the real, physical body, the synthetic unity of its parts, and the permanence and maximum speed of its motion; an epistemological meaning if it denotes the complete, exact, and true representation (perception-description) of the real, physical body; an ethical meaning if it denotes the physical body's supreme good; and an ontotheological meaning if it signifies the physical body's being, which we call *first* being or *supreme* being.

The Physical Meaning of the Real 1

*A*s the physical limit of accelerating expansion and progression, the real 1 designates the real, physical body defined as the limit and total sum of an infinite series of sensible parts a_n of increasing size and speed. We divide the real, physical body into the physical universe constituting the maximum limit 1 of our accelerating, sensible universe a_n and into the universal mind constituting the maximum limit 1 of our accelerating individual brain a_n. Because the real, physical universe (object) is the greatest and highest body perceived exactly by the greatest and highest mind, the universal mind (subject), whose organ is the universal brain, we conclude that the physical universe is both a perceived object and a perceiving subject; in other words, the physical universe is a self-perceiving being that exists and which also has an exact and true perception of its existence. As a matter of fact, we can claim that the physical universe is the universal brain-mind that perceives and thinks the physical universe. Given the one-to-one correspondence between the physical universe and the universal brain, we can regard the physical universe as a place and organ of universal perception, which we call *universal brain,* and the universal brain as an *organic universe.*[1]

---

1. The one-to-one correspondence between the physical universe and the

The Meaning of Acceleration

We have asserted that the *first cause*, which is simultaneously the *first end* and the *first principle* in the physical universe, is the physical universe itself numbered by the real 1 and governed by the synthetic

universal brain enables us to define the (inertial) mass of the physical universe in terms of the number of its contained cells. If the maximum mass of the physical universe of radius 10^{26} m is of the order of 10^{51} kg, and if the mass of the individual brain cell of radius 10^{-4} m is of the order of 10^{-9} kg, then the physical universe's number of cells is $10^{51}/10^{-9} = 10^{60}$.

Let us assume a law of matter, which is analogous to the square law of force obeyed by repulsive gravity (see formula 3.19, p. 80). This law states that the body's number of parts (cells, particles) or amount of inertial mass varies in proportion to the square of its distance d from a given point a:

1.
$$m_i = kd^2$$

where k is a constant of proportionality.

Consider an inertial mass m_i at the point a. Now surround it by a two-dimensional spherical surface of radius r, which is simultaneously the distance d from the center and starting point a. Because the spherical surface area is proportional to the square of its radius, the above relation suggests that it is the area of the surface that fixes the mass concentrated within a spherical region of space and not its volume. We can regard, then, the above relation as the *principle of self-creation and self-propagation of matter.*

Because matter (intensity or concentration) and space (extension) are equivalent, they are reciprocally dependent and caused and therefore free of external dependence and cause. Accordingly, space (manifested as vacuum energy, repulsive gravity, extended matter, or inertial part of matter) generates matter and matter generates space. We are then led to conclude that ultimately the real, physical body composed of matter and space needs no artificial, external cause—neither a violent and transcendent cosmic singularity (big bang) nor a fury God—to explain and create its existence: that is, its matter and space. It follows that the real, physical body is free of external cause because it is continuously self-caused. By *self-causality* (immanent causality—free or spontaneous causality) we mean the faculty of being both cause and effect of itself. For example, if one of the parts of the physical body—say, matter or space—is regarded arbitrarily as a cause, then the contrary part is an effect. On the whole, the physical body is both cause and effect and hence self-created.

principle of equivalence. If the *first cause* of being possessing extension and motion were not the extended and rotating physical universe itself but rather something absolutely different and separated from the physical

Let us come now to $k = m_i /d^2$, the constant of proportionality linking inertial mass to the surface area of a spherical volume of space of radius d. This constant determines the rate of continuous creation of mass: how much mass is created by a spherical region of space per unit of time—say, every 10^{10} (10 billion) years with respect to the universe's horizontal axis xx or 1.5×10^{10} (15 billion) years with respect to the universe's diagonal axis zz. We take as initial mass the smallest organic mass 10^{-9} kg, which is roughly the inertial mass of the brain cell of radius 10^{-4} m, and as surface area the area of a three-dimensional spherical region of space of radius 10^{-4} m. The creation rate is the following: 1 cell of mass 10^{-9} kg is created by a surface area of $(10^{-4})^2 = 10^{-8}$ m every 10^{10} years: $k = 1/10^{-8}$ or $k = 10^{-9}/10^{-8}$. We can likewise say that 1 cell of mass 10^{-9} kg belongs to a spherical region of space of radius 10^{-4} m. If the radius of the growing sphere is maximum according to extension—say, 10^{26} m—then the sphere's surface area depending on its square radius is maximum and its total created inertial mass or number of cells is maximum:

2. $\qquad m_i = 10^{-9}/(10^{-4})^2 \times (10^{26})^2 = 10^{51}$ kg $\quad$ or $\quad 10^{51}/10^{-9} = 10^{60}$ cells.

Here we see that the maximum surface area of a sphere depending on its maximum square radius—say, $(10^{26})^2 = 10^{52}$ m—creates, say, every 10^{10} years the greatest mass 10^{51} kg, or the greatest number 10^{60} of cells, which is the inertial mass of the greatest and heaviest physical body thought of as the cell of cells: that is, as the supreme cell. It also shows the converse, that the greatest mass or the greatest number of cells creates per unit of time the greatest surface area belonging to the sphere of the greatest and heaviest physical body.

Taking into consideration the above, we can think of the physical universe as a cosmic (universal) brain composed of 10^{60} cells immediately communicating between them via the outer space-time singularity of zero radius occurring in dimension zero.

Because maximum magnitude according to extension is the finite magnitude, which is simultaneously an infinite magnitude, it follows that the finite mass 10^{51} kg, defined as the greatest limit of an unlimited sequence of increasing masses, is simultaneously an infinite mass—that is, the inertial mass of *everything*: 10^{51} kg $= \infty$.

universe—namely, a simple, absolutely immobile and extensionless point (a *transcendent* cosmic singularity occurring in dimension zero and before the physical universe)—then how is it possible for this simple point without extension and motion to impart the properties of extension and motion to the physical universe? How is it possible to

If the distance d, regarded simultaneously as the radius r of the growing sphere, is a maximum according to division—say, 10^{-34} m—then its inertial mass depending on the surface area of the growing sphere is equally a maximum according to division:

3. $$m_i = 10^{-9}/(10^{-4})^2 \times (10^{-34})^2 = 10^{-69} \text{ kg.}$$

Here we see that the surface area of a sphere depending on its maximum square radius according to division—say $(10^{-34})^2 = 10^{-68}$ m—creates say, every 10^{10} years the smallest mass 10^{-69} kg, which is the inertial mass of the smallest and lightest physical body: that is, the mass of the elementary particle or *first particle*. It also shows the converse, that the smallest physical mass creates every 10^{10} years the smallest surface area belonging to the sphere of the smallest and lightest physical body.

Because maximum magnitude according to division is the finite magnitude, which is simultaneously a zero magnitude, it follows that the finite mass 10^{-69} kg, defined as the smallest limit of an unlimited sequence of decreasing masses, is simultaneously a zero mass—that is, the mass of *nothing* presented, depending on the case, as a soul, a spiritual or intellectual point (a noumenon), a fire point or photon, an elementary particle, a space-time singularity, an information bit: 10^{-69} kg = 0. If we divide the mass of the greatest physical body, which is the real whole or universe containing everything, by the mass of the smallest physical body, which is the part containing nothing and contained in everything, we obtain the number of smallest and lightest parts (nothings, souls, intellectual points, photons, elementary particles, cosmic singularities, information bits) that the greatest and heaviest body has: $10^{51}/10^{-69} = 10^{120}$. If each smallest undivided part (individual) is a complex universe divided into two coexisting parts, and if there are $n = 10^{120}$ parts, then the sum total of all parts stocked in the physical universe is 2^n, where $n = 10^{120}$. Because this finite, dimensionless number is a maximum, it is simultaneously an infinite number computing the universe's infinite memory power and material wealth: We write, therefore: $n = 10^{120} = \infty$.

have motion between absolutely different and unequal things, between the physical universe possessing extension and motion and the simple point of zero extension and zero motion?

Taking into consideration formulae 2 and 3, we conclude that in total, the maximum inertial mass of the physical universe is simultaneously the mass of *everything* (of the greatest and heaviest physical body) with respect to extension and the mass of *nothing* (of the lightest and smallest physical body) with respect to division. This *massive nothing* has the greatest, finite mass density if we divide its finite mass by its finite volume and obtain an order of magnitude equal to $10^{-69}/(10^{-34})^3 = 10^{30}$ gm/cm^3 (grams per cubic centimeter); we consider in turn this greatest mass density as the finite magnitude of infinite density: 10^{30} gm/cm^3 = ∞.

The inverse occurs with *massive everything*, which has the smallest mass density, if we divide its finite mass by its finite volume. Its order of magnitude is then: $10^{51}/(10^{26})^3 = 10^{-30}$ gm/cm^3. We consider in turn this smallest mass density to be the finite magnitude of zero density: 10^{-30} gm/cm^3 = 0. The synthetic product of *dense nothing* and *denseless everything* gives the mass density $10^{30} \times 10^{-30} = 1$, which is equal to the density of water, and to the density of the cell whose order of magnitude is 10^{-9} kg/$(10^{-4}$ m$)^3 = 10^{0}$ gm/cm^3. We then conclude that globally the physical universe composed of 10^{60} cells has—as the cell of cells—the supreme cell, the density of water. In this way the cell and the physical universe can possess the infinite density of the dense nothing without losing their organic properties relating to water by virtue of possessing at the same time the zero density of the denseless everything that neutralizes the effect of infinite density on their living bodies.

This leads us to conclude that cell life (or the living cell) is not a local, particular event; cell life is not an accident, a fleeting possibility, or a miracle occurring uniquely at the center *a* of the physical universe, there where organic life resides. Rather, cell life is a universal event and a universal principle occurring everywhere, under all conditions of infinite and zero densities, at all times. Indeed, if at any point of the real, physical universe the mass density of the universe is the real $1 = ∞ \times 0$ (the density of water resolved into infinite and zero densities), then necessarily any point of the physical universe must be an equilibrium place of cell life. We call this the *omnipresence* and *omnipuissance*

Because continuous motion is possible when unequal things are equal and continuous, we conclude the following: Only if the *first cause* is both a point without extension and motion and a line with extension and motion—that is, a *complex point line* or *extended point* representing one-dimensionally the physical universe—then the *first cause* has the power to impart the properties of extension and motion to the physical universe. Now, a physical universe, which is simultaneously the cause and the effect of its extended being and motion, is governed by the principle of self-causality (free causality or immanent causality) and self-motion (free or spontaneous motion).[2]

of unconditioned cell life (living cell) occurring everywhere, at all times, under all conditions—at the center *a* and thus inside the cold, infinitely extended universe of zero density and zero energy and on the limiting-circumference *b* and thus outside and beyond the universe in its hot, extensionless cosmic singularity of infinite density and infinite energy.

2. An immanent principle of the physical universe is reflexive order (self-order; free or spontaneous order), stating that the physical universe is anything that is numbered by the real 1 and has the property of being both before and after itself:

1. $$1 = (1 < 1)(1 < 1) \rightarrow 1 < 1.$$

If the sign < (the sign of inequality and temporal order) denotes the relation of implication, and if the left, antecedent part of the implication denotes the *cause* and the right, consequent part of the implication denotes the *effect*, then reflexive order becomes the principle of self-causality (immanent causality, free or spontaneous causality), which states that the physical universe is anything having the property of being both the cause and the effect of itself.

If the left, antecedent part of the implication signifies the *observing subject* and the right, consequent part of the implication signifies the *observed object*, then we can define the physical universe as anything that is both an observing subject and an observed object.

The synthetic principle of reflexive order (presented as a principle of self-causality, self-motion, self-perception, self-containment) has its historical origin in the Anaxagorian cosmological doctrine of the real, physical universe. According to Anaxagoras, the physical universe is a body rotating permanently

If self-causality and self-motion are immanent principles of the timeless, physical universe, then a physical universe existing and moving by an *absolutely external, transcendent,* and *simple cause* (which may be Aristotle's unmoved mover, the fiery God of monotheism, or the violent big bang of the evolutionary cosmological model) is an improper, physical universe. In reality, it is a time-conditioned, sensible universe assimilated to the observable part of the physical universe. Thus, the time-conditioned sensible universe has a *first cause,* which is a *transcendent, simple point* without extension, motion, and being occurring uniquely outside of and before the sensible universe in dimension zero. However, insofar as this *simple, first cause* generates absurdities (for example, that of generating extension from zero extension, motion from immobility, line from point, being from not-being), in no way can it constitute a *real, first cause* of being endowed with extension and motion; in no way can it create, restore, sustain, and extend the extended being and its motion. Thus, the *real, first cause* of being with extension and motion is the complex physical universe itself occurring both within and beyond

at a maximum speed. Its continuous rotation is initiated and controlled by the universal mind (νοῦς), which is infinite (has infinite being, infinite knowledge, and infinite force), self-ordered or self-governed (αὐτοκρατές), and corporeal.

We can derive the above principle of reflexive order from the synthetic principle of equivalence, which governs the physical universe and stipulates the equality of unequal things—for example, of the indefinitely varying part a_n with its constant whole 1.

Let us define the equivalence $a_n = 1$ as the product of opposite inequalities and temporal orders:

2. $$(a_n = 1) = (a_n < 1)(1 < a_n).$$

If we replace a_n with its equal 1 in $(a_n < 1)(1 < a_n)$, we can then define the equivalence $a_n = 1$ as reflexive inequality or reflexive order:

3. $$(a_n = 1) = (1 < 1)(1 < 1) \rightarrow 1 < 1.$$

This equation shows that the principle of reflexive order is a variant of the synthetic equivalence principle, which governs the real, physical universe.

the time-conditioned, sensible universe a_n and hence on the greatest circle (limiting-circle) b, which is an infinitely curved straight line equivalent to Plato's *point line* (ἄτομος γραμμή).

The Epistemological Meaning of the Real 1

And when these things have been thus separated,
we must know that all things are neither more nor less;
for it is not possible that there should be more than all,
but all things are always equal.

—Anaxagoras

We have claimed that from the physical point of view, the *end* of our accelerating sensible universe a_n, of our accelerating galaxies and individual brains is to attain the real 1, which signifies the timeless, physical universe rotating continuously at the maximum speed of light $c = 1 = 0 \times \infty$ and governed by the synthetic equivalence principle.

If, according to the synthetic principle of reflexive order assigning to the physical whole the power of self-perception, anything being a physical universe (subject) has simultaneously an exact and universal perception of its own existence as a physical universe (object), then the attainment of the state of being a physical universe involves the attainment of a complete and universal perception of the real, physical universe. It follows from the epistemological perspective that the *end* of inductive growth, of the accelerating analytic science and individual brain perceiving incompletely or partially the physical universe, is the complete and universal perception of the physical universe by the universal mind or universal brain rotating continuously at the maximum speed of light. Because it moves at the maximum speed of light, the universal brain perceives the physical whole 1 through physical light, *the first light*, and hence as it *really is* (τό καθαυτό): that is, universally or spherically from all senses—from inside as an accelerating part a_n and from outside as a constant, limiting-point b—a cosmic singularity transforming the accelerating part a_n into the real, infinite whole 1 rotating continuously at the maximum speed of light.

We have shown that in order to obtain an exact and universal perception of the physical universe, of the real, infinite whole 1, our low-energy individual brain must become what it is already and has been for eternity: a universal brain elevated to the level of highest energy, unity, and universality, which is the outer, convex part (the cosmic singularity part) of the universe's limiting-circumference b. This inductive elevation of energy is accomplished through the maximum contraction or acceleration of the brain's cells becoming themselves *organic singularities*. Each organic singularity or high-energy brain cell possesses a maximum energy content equal to the Planck energy, which is simultaneously the energy of the infinite part of light traveling via the universe's cosmic singularity with infinite speed or zero wavelength and thus with infinite frequency. This infinite part of light ensures that the high-energy universal brain detects the entire physical light propagated spherically everywhere, in all senses, at all frequencies (at infinite and zero frequencies) and immediately carrying information about the entire physical universe, its inner infinitely extended part and its outer extensionless part, its past and its future. The ideal faculty for perceiving the physical universe completely and spherically in all senses, from all parts, we called infinite, synthetic, universal sensibility. This universal sensibility is the faculty that permits the universal brain or universal mind to have a complete and universal perception of the real, physical whole 1, to have a complete and universal communication with any of its parts, and hence to be a member of nature's universal community.

Because the universal brain perceives the physical universe completely and spherically in all senses, the spherical or universal perception of the physical universe does not alter the content of the physical universe. This means that the physical universe is the same before and after universal perception. It means that the physical universe verifies the first law of thermodynamics, which is, in its ontological expression, the principle of conservation and nonalteration of the physical universe, and, in its logical expression, the synthetic principle of reflexive order stating that the physical universe—the real 1—is the same before and after itself: $1 = 1 < 1$. It also means that universal perception is a forceless and timeless action!

The infinite compression of the *infinitely many* things into the limiting *one* thing constituting the real, physical body shows that the real 1, regarded as the *unity* of the *infinitely many*, is also a principle of the infinite complexity of the physical body and its contained parts. Indeed, if we decompose (divide) the real, infinite whole 1 into contrary (opposite and coexisting) parts a and b such that 1 is their synthetic product, or constant ratio, and hence their synthetic unity (proportionality):

$$5.1 \qquad 1 = ab, \qquad \text{or} \qquad 1 = a/b \rightarrow a = b,$$

and if we assign to each part of the equality $a = b$ the two parts a and b an infinite number of times, we then obtain an infinite geometric progression of parts of increasing complexity whose limit is the real 1 $= \infty \times 0$.

$$5.2 \qquad 1 = (a = b) \rightarrow (aa)(ab)(ba)(bb) \rightarrow (aaa)(aab)(aba)(abb)(baa)(bab)$$
$$(bba)(bbb) \rightarrow (aaaa)(aaab)(aaba)(aabb)(abaa)(abab)(abba)(abbb)(baaa)$$
$$(baab)(baba)(babb)(bbaa)(bbab)(bbba)(bbbb) \rightarrow \ldots = 1 = \infty \times 0.^3$$

3. Because we assume that every part of the real, infinite whole $1 = (a = b)$ is equal to the real whole 1, every part has the synthetic property of the real

The real 1 computes through its number ∞, where ∞ = 2^∞, and ∞ = 10^{120}, the infinite multiplicity of parts stocked in the above matrix and determining the physical universe's maximum or infinite power and wealth (see chapter 5, note 1, p. 173). On the other hand, the real 1 computes through its number 1 the above infinite multiplicity of parts as one thing—that is, as the sum *total* of all parts. Finally, the real 1 computes through its number 0 the universe's nothingness, the absence of parts, by virtue of being a cosmic singularity—a limiting-point *b* in dimension zero, which has no parts.

Can the real 1 = (*a* = *b*) from which we unfold the infinite geometric progression of parts (*aa*)(*ab*)(*ba*)(*bb*)… describe completely the physical universe and can it answer all questions posed by our finite, analytic science and individual brain to the physical universe? For example, can this real 1 answer the question of the shape of the physical universe, of whether the physical universe is finite or infinite with respect to extent—for instance, with respect to one dimension?

whole 1, which is the power of comprising at one instant contrary (opposite and coexisting) parts *a* and *b*. We thus assign to the unique part *a* the two parts *a* and *b* in order to obtain *aa* and *ab* and to the unique part *b* the two parts *a* and *b* in order to obtain *ba* and *bb*. We now assign to the unique part *aa* the two parts *a* and *b* in order to obtain *aaa* and *aab*; to the unique part *ab* the two parts *a* and *b* in order to obtain *aba* and *abb*; to the unique part *ba* the two parts *a* and *b* in order to obtain *baa* and *bab*; to the unique part *bb* the two parts *a* and *b* in order to obtain *bba* and *bbb*. Continuing in this way an infinite number of times, we obtain an unlimited geometric progression of parts of increasing complexity.

The founding principle of the above assumption is the whole–part equivalence principle. This is a particular variant of the physical universe's synthetic equivalence principle, the *first principle* that governs the universe's greatest sphere and stipulates the equality of unequal things—for example, the complex whole and the simple part. Accordingly, because the part is equivalent to the whole, the part has the synthetic properties of the whole—namely, the power to comprise contrary parts *a* and *b* by virtue of being extended and divided.

Our rationalistic assumption is that this complex, real 1 defined as the synthetic unity of opposite answers (parts) *a* and *b* is, indeed, the real, definite, and complete answer, the *first answer*, to this question and to all questions of the general form: Is 1 *a* or *b*? [4] Accordingly,

4. An empiricist (an Aristotelian or a Kantian), who assumes that the physical universe numbered by the real 1 is simple and determinate, verifying analytic principles of being (analytic ontology of the being), will assert that the physical universe must give a simple (univocal) and determinate answer to any simple question of the general form: Is 1 *a* or *b*? It follows that the impossibility of the physical universe to give such a simple, determinate answer to any question indicates our ignorance of the simple nature of the real, physical universe, where 1 is either *a* or *b*. Now, the contrary happens with a rationalist, who considers the impossibility of giving a simple answer to any question as in fact the complete answer to the question! Indeed, a rationalist (a Platonist), who assumes that the physical universe, the real 1, is complex and indeterminate, verifying synthetic principles of being (synthetic ontology of the being), will assert that the physical universe must give a complex (equivocal) answer to any posed question, and that the impossibility of giving a simple (univocal) answer—far from indicating our ignorance—reveals our complete knowledge of the complex, indeterminate nature of the physical universe, of the real 1, where 1 is both *a* and *b* and neither *a* nor *b*.

An empiricist is the low-energy individual observer at the low center *a* who perceives through his or her particular sensibility only the sensible part of physical light traveling at the finite speed $c = 1$. This sensible light reveals the physical whole 1 uniquely from inside, *as if* it were an indefinitely accelerating simple, sensible part a_n, which verifies analytic principles of being (a case of pereptual illusion). He or she then takes this indefinitely accelerating simple, sensible part a_n *as if* it were the real, physical whole and asserts that because the real, physical whole is a simple universe, a finite whole, it must give a simple, determinate answer to a simple question (a case of ontological and epistemological illusion). In reality, however, the real, physical whole, which is a complex universe, a real, infinite whole 1 verifying synthetic principles of being, constantly contradicts the assertion of the empiricist by giving complex, indeterminate answers to simple questions. The contradiction is resolved as soon as we replace the empiricist by the rationalist, according to whom the

if the real 1 is the one-dimensional physical universe governed by the synthetic equivalence principle, and if *a* designates the infinite answer according to which the one-dimensional physical universe is open and infinite described by a straight (or hyperbolic) line, *b* designates the finite answer according to which the one-dimensional physical universe is closed and finite described by a circular line or point, and (=) designates unity and simultaneity, then the complex, real 1 = (*a* = *b*) is the synthetic unity and simultaneity of the two answers *a* and *b*. Because both partial answers *a* and *b* have equal claim to truth, their simultaneous existence tells us that the exact and true description of the one-dimensional physical universe is the composition (unity, product, contact) of the open, infinite straight line, with the closed, finite circle, which Plato called the *undivided line* (point line, ἄτομος γραμμή).

Indeed, seen from inside, from its center *a*, our one-dimensional physical universe is an infinite straight line of zero curvature, whereas seen from outside, from its outer limiting-circumference *b*, our one-dimensional physical universe is a finite, circular line or point of infinite curvature.[5] Another way of describing the composition of the infinite

complex, indeterminate answers given permanently by the physical universe reveal its real, complex, indeterminate nature, and hence our exact and complete knowledge of this complex, indeterminate nature numbered by the real, infinite whole 1.

5. An alternative way of deriving the complex nature of the physical universe's dimension from its real number 1 = (*a* = *b*) is this: Let *a* designate the average matter density of the universe and *b* designate the critical matter density of the universe whose order of magnitude is: 10^{51} kg/$(10^{26}$ m$)^3$ = 10^{-30} gm/cm^3— that is, the inverse of the physical universe's relative extension 10^{26} m/10^{-4} m = 10^{30}. In this case, the complex, real 1 = (*a* = *b*), defined as the product of two inequalities:

$$1 = (a = b) \text{ if and only if } (a < b)(b < a),$$

states: The physical universe 1 has its average matter density *a* equal to the critical matter density *b* if, and only if, its average matter density *a* is both less than and greater than the critical matter density *b*.

and the finite is through the complex, limiting circle of magnitude 1 whose inner, concave part is equivalent to the open, infinite straight (or hyperbolic) line and whose outer, convex part is equivalent to the closed, finite circle or point. Now, if we negate both answers a and b of the one-dimensional physical universe numbered by the real $1 = (a = b)$, we then obtain the equivalent structure $1 = (a' = b') = (a + b)'$, stating that the one-dimensional physical universe is governed by the synthetic principle of included third: that according to this principle, both answers a and b are false, that neither answer a nor answer b is true. We then predict or conclude that the one-dimensional physical universe is neither an infinite straight line of zero curvature nor a finite circular line or point of infinite curvature and therefore that the physical universe is indeterminate and impartial, free of the constraining partiality of the answer.

The two answers a and b given by the real $1 = (a = b)$ are divided or multiplied into four answers if we correlate them in all possible ways relative to the universe's two dimensions, the horizontal x and the vertical y (see table 5.1):

If $a < b$ determines an open and infinite universe with respect to extension where its average matter density a is less than the critical matter density b and $b < a$ determines a closed and finite universe with respect to extension where its average matter density a is greater than the critical density b, then the above formula states that the average density a is equal to the critical density b if, and only if, the physical universe is both open and closed, infinite and finite with respect to extension. This means that if empirical cosmology shows (through cosmic microwave data, for example) that the average density a is equal to the critical density b, then we can inductively predict or conclude that the real, physical universe is both open and closed, infinite and finite with respect to extension: that in a general manner, any dimension of the physical universe has a complex shape by virtue of being an infinite straight line of zero curvature when seen from inside and a finite, circular line (or point) of infinite curvature when seen from outside.

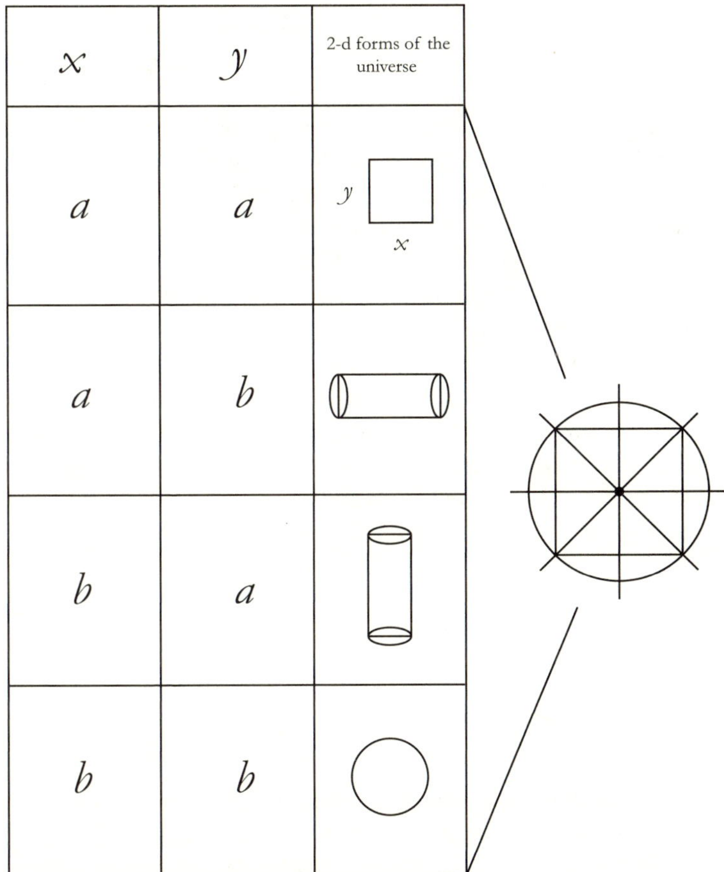

x	y	2-d forms of the universe
a	a	
a	b	
b	a	
b	b	

Table 5.1. Here we see the totality of ways of correlating *a*, designating the infinite, straight (or hyperbolic) line, and *b*, designating the finite, circular line or point, with the physical universe's two dimensions, the horizontal dimension *x* and the vertical dimension *y*. The derived correlations constitute the four partial answers that the two-dimensional universe gives to this simple question: Is the universe finite or infinite with respect to extent—with respect to two dimensions? Because all four answers are equally true, their coexistence tells us that the true and exact description of the two-dimensional physical universe is the synthesis (composition) of the finite circle (point) and the inscribed infinite square.

- The answer (*aa*) describes the two-dimensional physical universe as being open and infinite and of zero curvature relative to its horizontal and vertical dimensions *x* and *y*—that is, as being a Euclidean square.
- The answer (*ab*) describes the two-dimensional physical universe as being open and infinite and of zero curvature relative to its horizontal dimension *x*, but closed and finite or pointlike and of infinite curvature relative to its vertical dimension—that is, as being the surface of a horizontal cylinder.
- The answer (*ba*) describes the two-dimensional physical universe as being closed and finite (or pointlike) and of infinite curvature relative to its horizontal dimension *x* but open and infinite and of zero curvature relative to its vertical dimension *y*, and hence as being the surface of a vertical cylinder.
- The answer (*bb*) describes the two-dimensional physical universe as being closed and finite (or pointlike) and of infinite curvature relative to its horizontal and vertical dimensions *x* and *y*, and hence as being the surface of a sphere.

Continuing in this way, we can geometrically increase the answers and shapes of the physical universe as we geometrically increase the universe's dimensions in function of the one-to-one correspondence relation $2^{n-1} = 2^n$, where 2^{n-1} is the number of dimensions and 2^n is the number of answers or shapes.

Because each answer is a part of reality, the sum total of all answers constitutes the maximum reality (*ens realissimum*) of the physical universe. The sum total of all answers given by the physical universe reveals nature's *omnipotence* to actualize any possible shape in the physical universe; it reveals equally the maximum wealth of shapes that the physical universe stocks in itself. The maximum storage capacity of the physical universe of mass 10^{51} kg is 10^{60} times the brain cell's maximum storage capacity of mass 10^{-9} kg. Accordingly, if the cell's maximum storage capacity at the state of high energy is 2^n, where $n = 10^{60}$, then the maximum wealth and power of the physical universe is $2^{n \times n}$, where $n \times n = 10^{60} \times 10^{60} = 10^{120}$ (see chapter 4, note 20, p. 155, and chapter 5, p. 181). This maximum number is the finite computation of the physical universe's infinite wealth of shapes.

Compressing this infinite wealth of shapes into two shapes given by the two answers *a* and *b*, we conclude the following general statement, which permits us to describe exactly the shape of the one-dimensional physical universe independent of the infinite geometric progression of its shapes and answers: Any dimension of the physical universe numbered by the real 1 has a complex and indeterminate shape, as it emerges as a constant ratio of zero and infinite curvatures and therefore is both infinite and finite and neither infinite nor finite. A complex and indeterminate universe, which is with respect to one dimension both and neither infinite and finite, is a balanced and permanent universe free of time.

Another way of developing the real $1 = (a = b)$, which is the physical universe's *first principle*, the synthetic equivalence principle, is:

5.3 $$1 = (a = b) \rightarrow (ab = ba) \rightarrow (ab < ba)(ba < ab),$$

where *a* and *b* are any two different parts of the physical universe and the sign of inequality < designates the relation of temporal and causal order—of generation and implication between the different parts. The left, antecedent part of the implication we call *cause*; the right, consequent part of the implication we call *effect*. After having resolved (divided) the equality $ab = ba$ into opposite and coexisting inequalities and causal orders, such as $(ab < ba)(ba < ab)$, we obtain the principle of indeterminate causality, which affirms that everything—namely *a* and *b*—implies and is implied by everything. Indeterminate causality is a variation of the synthetic equivalence principle: the principle grounding continuous motion between the unequal parts of the universe.

As we have argued several times, continuous and immediate communication between the unequal parts $a < b$ is real and complete if, and only if, the unequal parts are equal. However, the equality of unequal parts transforms the unequal, determinate parts $a < b$ into equal, indeterminate wholes $ab = ba$ in continuous, immediate communication, where anything implies and is implied by anything.[6]

---

6. Different ways of developing the synthetic equivalence principle $1 = (a = b)$ produce different expressions of its correlative principle of indeterminate

Thus, if continuous motion is a real feature of the physical universe, then the physical universe and its parts are equivalent, infinite wholes governed by the synthetic principle of indeterminate causality, which is the end of determinate causality and its time-asymmetric structure.[7]

causality. Thus, if we develop $1 = (a = b)$ as follows:

1. $\qquad 1 = (a = b) \rightarrow (a = ab) \rightarrow (a = b'b) \rightarrow (a < b'b)(b'b < a),$

we obtain the synthetic principle of indeterminate causality, affirming that one cause a implies different effects b' and b (or an infinity of effects) at the same time in conformity with the indeterminate structure $a < b'b$, and that one effect a is implied by different causes b' and b (or by an infinity of causes) at the same time in conformity with the indeterminate structure $b'b < a$. We conclude, then, that one thing implies or is implied by everything or anything.

Let a and b be propositions (or theories) about an extended thing and each true proposition be equal to the falsity of its counter-proposition: $a = b'$ and $b = a'$. If we replace a with b' in $(a = ab)$, we obtain the indeterminate causality:

$$(a = b'b) \rightarrow (a < b'b)(b'b < a),$$

where a true proposition (or theory) a implies or is implied by different propositions b' and b at the same time. For example, the same true cosmological theory a proves both opposite propositions b' and b about the physical universe, where b states that the physical universe is open and infinite with respect to extent and b' states that it is false that the physical universe is open and infinite with respect to extent. This is equivalent to saying that the same true cosmological theory a, which is a complex theory reflecting the complex and indeterminate nature of the extended real thing—the physical universe—simultaneously proves and disproves proposition b and that proposition b is both true and false (see also the synthetic universal proposition u represented geometrically by the diagonal or circular line in chapter 3, note 20, p. 84–86).

7. Determinate causality (external, analytic causality), stipulating that one thing a implies either b or b' in conformity with the analytic, determinate structure $(a < b) + (a < b')$ or that one thing a is implied by either b or b' in conformity with the analytic, determinate structure $(b < a) + (b' < a)$, is not a feature of the real thing—the physical universe. Rather, it is a feature of the time-conditioned observable part a_n of the physical universe, which we called sensible universe, accelerating toward its perfection, the timeless, physical universe, numbered and governed by the real 1. As a feature of the incomplete, sensible universe, determinate causality is an observable, Euclidean part of the totality of causal orders

Far from announcing the chaotic collapse of both the physical universe and its continuous motion, indeterminate causality expresses

that determines the entire sphere of the physical universe and renders its causality globally indeterminate. For example, the determinate, causal order $a < b$ is an observable part of the totality of causal orders $a < b'b = (a < b)(a < b')$, while the converse determinate causal order $b < a$ is an observable part of the totality of causal orders $bb' < a = (b < a)(b' < a)$.

If the totality of causal orders, and thus indeterminate causality, determines the real, physical universe at any time, it follows that a particular causal order (for example, $a < b$), selected arbitrarily by our particular senses at a definite time, does not determine the real, physical universe itself, but instead determines the incomplete, analytic, and time-conditioned manner with which our low-energy individual brain perceives the physical universe. Analytic, determinate causality is, then, an apparent or improper principle of the physical universe. To put it another way, determinate causality is a perceptual illusion of our low-energy individual brain and its mental faculty of finite, particular sensibility. In the perfectly uniform, timeless, physical universe, where time is the $n + 1$ dimension of n-space, determinate causality involving time asymmetry disappears and indeterminate causality emerges as an immanent and real principle of the physical universe, of the real 1.

Thus, if through our low-energy individual brain and its mental faculty of finite, particular sensibility detecting only a determinate part of physical light we experience the universe as a discontinuous multiplicity of conflicting parts called sensible universe, we are then forced to use the analytic principle of determinate causality to explain why the true theory a, which is a simple theory reflecting the simple and determinate nature of the sensible part of the physical whole, implies proposition b rather than proposition b' or proposition b' rather than proposition b. Per contra, if through our high-energy universal brain and its mental faculty of infinite, universal sensibility detecting the entire physical light we experience the universe and any of its parts as a continuous multiplicity of unified parts called real, physical universe, then determinate causality to explain the indeterminate, physical whole is needless. Indeed, in the real, physical universe anything implies and is implied by anything at once in conformity with the synthetic principles of equivalence and indeterminate causality and as a consequence of the unifying action of the physical universe's cosmic singularity.

its infinite power and absolute freedom from external cause; in this sense, it is a principle of free causality. Indeed, if anything implies and is implied by anything in the timeless, physical universe, then everything is free of an external, determinate cause and obeys the synthetic principle of free (spontaneous) causality or self-causality (immanent causality). According to this latter principle, everything—the physical universe and any of its parts—is both a cause and an effect of its continuous motion and requires neither asymmetric time nor an external, determinate cause (constraining force), such as a *transcendent* and *simple, first cause,* to generate, explain, and sustain its extended being and motion.

Free causality, which is a variant of reflexive order (self-order), assigns to the real, physical whole 1 the infinite powers of self-containment, self-government, self-knowledge, and self-motion.

We have argued in this chapter that whenever our analytic science and our resting individual brain ask whether the physical universe of radius $ab = 1$ is a or b, the physical universe answers this simple question in terms of its first principle, the real $1 = (a = b)$, thought of as the synthetic equivalence principle. This complex and indeterminate principle states that the physical universe of radius $ab = 1$ is the synthetic unity and product of different parts and answers a and b. It also states that the real thing—the physical universe—emerges as the constant, balanced, and just ratio of opposite answers: $1 = a/b$.

Our individual brain asks the physical universe simple questions, because in its state of rest it experiences the physical universe and its emitted physical light uniquely from inside as an indefinitely accelerating simple sensible part revealed by a simple sensible light of definite wavelength λ traveling through space-time in a unique sense, at the unique, finite speed $c = 1$, and verifying analytic principles of being. The individual brain considers next this inner, simple, sensible part verifying analytic principles of being *as if* it were the real, physical universe itself giving a unique or simple answer—namely, either the answer *a* or the answer *b*—to this simple cosmological question: Is the physical universe with respect to extent *a* or *b*? But the impossibility of deciding intellectually and empirically one of these answers indicates either the ontological impossibility of the physical universe, that the physical universe does not exist, or its epistemological impossibility, that if the physical universe exists, it is impossible to know its nature, which is assumed to be simple and hence to be with respect to extent either *a* or *b*.

From the low, analytic point of view, the physical universe's *first principle*, the real $1 = (a = b)$, cannot constitute a complete answer to the above simple, cosmological question (in fact, because of its complex and indeterminate nature, the real 1 constitutes the impossibility of giving a simple and determinate answer to the question). Nevertheless, from the high, synthetic point of view, the real $1 = (a = b)$ constitutes the true and complete answer to the cosmological question—nay, the ultimate resolution of all questions—bringing serenity, satisfaction, and rest without destroying life and motion.

Thus, if we elevate our low-energy individual brain and its corresponding mental faculty of particular sensibility to the level of highest unity (to the cosmic singularity level in dimension zero) in order to obtain a spherical or universal perception of the physical universe, we then perceive the *real* form of the physical universe, whose one-dimensional figure is the undivided line (point line) or the infinitely extended point (line point), and whose two-dimensional figure is the infinite square inscribed to the finite circle. Seen from inside, from the low center *a*, the physical universe is an infinite Euclidean square proving the infinite answer *a*. Seen from outside, from the high limiting-circumference *b*, the physical universe is a limiting circle of zero radius, an infinitely curved cosmic singularity closing the infinitely extended Euclidean square and proving the finite answer *b*. The universal perception of the physical universe from all its inner and outer parts, its forward and backward senses, enables us to construct the perfectly true and comprehensive cosmological theory about the physical universe that proves rationally and empirically both answers $a = b'$ and $b = a'$, and hence neither answer b nor answer a: $ab = b'a' = (b + a)'$.

Thus, the perfectly comprehensive cosmological theory realizing the highest, synthetic unity of different but equally true (and false) answers does not tell us *why* the physical universe is at one time one thing rather than another in conformity with the analytic principle of determinate causality (Leibniz's principle of sufficient reason), but rather *what is* (τί ἐστιν) the real, physical universe at all times independent of our particular senses and their determinate causality, and in conformity with the synthetic principle of indeterminate causality.

Because, according to the perfect cosmological theory, the physical universe numbered and governed by the real $1 = (a = b)$ is eternal, is equal in its parts and answers, and is free and just, there is no sufficient reason to choose one part over the other, to choose answer a over answer b or answer b over answer a. In fact, nothing in the perfectly uniform and just universe is one thing rather than another, and this is what constitutes the infinite nature of the timeless, externally causeless, and therefore self-caused physical universe.

The Ethical and Ontotheological
Meanings of the Real 1

*T*he real 1 = (a = b), stipulating the equality of unequal parts, of any part a with any other part b of the physical universe, is essentially a principle of *universal* equality, which we call the *perfect equivalence principle*. We qualify as *supreme good* the perfect equivalence principle and its derived synthetic principles of self-causality (reflexive order) and indeterminate causality; they govern the real, physical universe (object) and the universal brain (subject) knowing completely its real self and the real, physical universe endowed with permanent life and continuous motion.

From the cosmological point of view, the synthetic principle of universal equality leads us to the principle of perfect uniformity, according to which the physical universe (its space-time) is the same everywhere (homogeneity), in all directions (isotropy), at all times (eternity). This sameness is possible because the physical universe's curvature, which is equal to the real 1 = $\infty \times 0$, is constant—that is, the curvature is the same at every point, in every direction, at all times. The perfect uniformity of the physical universe allows its matter-energy-information to be distributed equally, to be the same everywhere, in all directions, at all times, and thereby to determine in the physical universe a state of cosmological justice and balance necessary for the continuity of life and motion.

Relative to different domains of the world, the ideal principle of universal equality appears as the epistemological principle of highest reason (intelligence) and truth, as the aesthetic principle of supreme symmetry and beauty, as the physical principle of permanent life and continuous motion, as the *cosmico nexus* principle of universal love and immediate communication at a distance, and as the ethical principle of supreme good involving universal justice, universal happiness, and supreme freedom.

Along with these considerations, we can think of our accelerating sensible universe a_n as an infinite sequence of increasing equalities (or decreasing inequalities) striving to attain the real 1, the supreme good, which is the perfect equivalence principle (principle of universal equality), stipulating the equality of all parts of the real, physical universe.

If the supreme good involves freedom from evil and the source of all evil is the analytic principle of inequality and temporal order (the principle that breaks the unity and balance of the world, the continuity of its life and motion), then the supreme good is freedom from inequality and temporal order. The physical consequence of attaining the supreme good, the perfect equivalence principle, is the realization of the ideal state of permanent life and continuous motion, which is the feature of the physical universe—this being of beings called the *first being* numbered and governed by the real 1.

Let us call *religion* the unity of the progressing individual brain a_n with the real 1, the *first being* possessing immediately and effortlessly its supreme good, the perfect equivalence principle. At the limit 1, the low-energy individual brain is transformed into a high-energy universal brain possessing the synthetic faculty of infinite, universal sensibility. This is our individual brain's faculty of finite, particular sensibility elevated to the level of highest curvature, energy, and unity (the cosmic singularity of infinite curvature and infinite energy), unifying everything in dimension zero. The energetic elevation of our individual brain is accomplished by the maximum contraction or acceleration of its cells through the yogi method or through technology—the technology of the infinite whole. We have claimed that when the brain's energy content and speed are increased maximally, the brain experiences, through its mental faculty of infinite, universal sensibility, the supreme good, which is the perfect equivalence principle grounding the brain's continuous passage to the real 1. The attainment of the real 1 by the universal brain is equivalent to experiencing the divine state of the *first being*, regarded as the *first cause* and *infinite source* of highest intelligence and truth, of supreme justice and freedom, of universal communication and love, of permanent life and continuous motion.

By *permanent life* we mean continuous existence anywhere, in any direction, at any time, and under any condition. The permanently living being may appear to our low-energy individual brain, which is perceiving the living being uniquely from inside *as if* it were a time-conditioned living being artificially created by the external action of a *transcendent* demiurge, and as located uniquely at the low region of the center *a*. In reality, however, the permanently living being is a timeless *first being* unconditioned by an external cause. The permanently living being is governed by the perfect equivalence principle and its derived principle of self-causality. As an unconditioned being, the permanently living being occurs anywhere, at any time in the physical universe, under any condition: The unconditioned living being occurs at the low-cold region of the inner center *a* relative to the observer O and also at the high-hot region of the outer limiting-circumference *b*—the cosmic singularity region relative to the observer O′ (see fig. 4.7, p. 159).

Indeed, if according to the perfect equivalence principle the center a is equivalent to the limiting-circumference b, then the same living being at the inner center a is distributed spherically (uniformly) everywhere, in all directions, on the limiting-circumference b of the physical universe, whose outer part is the cosmic singularity region. It follows that the living being is not an improbable outcome of the physical universe but instead its very immanent nature. In fact, the self-contained physical universe is the permanently living being in continuous rotation and vibration.

By *universal communication*, we mean the power of continuous and immediate communication with anything, at any time, by passing through the physical universe's unifying singularity at the maximum speed of light $c = 1 = \infty \times 0$. The unifying cosmic singularity enables the unconditioned *first being* to communicate immediately with anything by being immediately present in everything and thus to possess the power of omniscience by omnipresence. In this way, the conflicting distinction and distance between the known object and the knowing subject is abolished.

Complete and direct perception of the physical object requires that the universal brain perceives the physical universe from all senses without loss of physical light, as it really *is* (τó καθαυτó), and hence from inside and outside as a real, infinite whole 1. This spherical perception is possible because the universal brain, rotating at the maximum speed of light $c = 1 = \infty \times 0$, is itself light, having its high-energy retinal cells detecting light directly. Direct perception of physical light means that there is no loss of physical light during the process of perception, that the physical universe and its emitted physical light are the same before and after perception. Direct perception also means that the sensible part a_n and the real 1 designating the intelligible, physical whole are equal. Thus, the physical universe defined as the sum total of all sensible parts is itself, in conformity with the whole–part equivalence principle, a sensible part.

Insofar as the physical universe is a member of its Euclidean series of sensible parts, it has the property of sensibility: The physical universe is an indefinitely accelerating sensible part represented one-dimensionally by the straight (or hyperbolic) line. Insofar as the physical universe occurs beyond its series of sensible parts, it possesses the property of intelligibility; it is an intelligible or ideal limiting point—a unifying cosmic singularity that the metaphysician calls the cosmic soul (Ψυχή), the cosmic intelligence (νούς) or spirit (πνεύμα). In total, the self-contained physical universe is, relative to contrary senses, both sensible and intelligible without absurdity.[8]

If by virtue of the transcendental principle of included third we negate both parts, the intelligible and sensible parts of the physical universe, we then obtain a real, infinite whole 1, which is neither intelligible nor sensible but rather a neutral, indeterminate, intermediate being occurring between its contrary parts and at the same time transcending them. This intermediate, and therefore synthetic, indeterminate real 1,

---

8. In an analogous manner, we assert that insofar as the physical universe is a member of the series of its parts or things, it has the properties of *thingness*. Insofar as the physical universe occurs beyond the series of its things, it has the property of *nothingness*. Ultimately, the physical universe is both one thing and nothing without contradiction or paradox. It follows that the physical universe is not one thing rather than nothing (which is an empirical fact requiring the causal explanation of *why* the universe is one thing rather than nothing in conformity with Leibniz's principle of sufficient reason), but instead both one thing and nothing, neither one thing nor nothing free of the partiality and artificiality of external causal explanation. Finally, if the property of the time-conditioned thing is age, we can equally assert that insofar as the physical universe is a member of the series of aging things, it has the property of age. Insofar as the physical universe occurs beyond the series of its aging things, it is ageless. On the whole, the self-contained physical universe, defined as the sum total of all ages, the age of ages, the *first age*, has simultaneously from inside an indefinitely increasing age and from outside a zero age, which transforms the indefinitely increasing age into an infinite age. The product of infinite age and zero age gives the finite maximum age equal to the real $1 = \infty \times 0$.

which is an intelligible object of our universal thought and a sensible object of our universal sense, is, according to Plato, the real object of mathematics. Thus, the *real end* of mathematical science is to realize empirically the continuous passage to the synthetic, indeterminate real 1, to perceive the real 1—the physical universe from all sides—to experience its supreme good, which is the perfect equivalence principle stabilizing its varying body, and to communicate immediately with all its spatiotemporal parts.

The inductive growth of mathematical science has meaning if, and only if, it helps the human being to experience its supreme good, which is the real 1—the physical universe governed by the perfect equivalence principle. Because the perfect equivalence principle affirms the end of inequality and temporal order, we can consider it as being esssentially a principle of freedom from corruption and evil. Indeed, temporal order, regarded as the analytic principle of irreflexive (heteronomous) order, generates the evil events of subordination, comparison, external causation, injustice, violence, war, and isolation that destroy the unity, energy, and balance of the world, its permanent life and continuous motion.

But what is the origin of inequality and of temporal order? According to our theory, it is our low-energy individual brain, which, by uniquely detecting the inner finite part of physical light traveling through space-time at the finite speed of light $c = 3 \times 10^8$ m/s $= 1$, perceives the real, physical whole 1 successively *as if* it were a time-conditioned accelerating part a_n composed of unequal, discontinuous parts, the individuals deprived of unity and verifying analytic principles of being. Because Aristotle regards inequality and temporal order as the founding principle of motion, what is significant for the discontinuous, sensible universe is its conservation. This Aristotelian assumption, however, is false![9] Using

---

9. Let F be the force of action and R the force of resistance acting on a body. According to Aristotle, the body moves if, and only if, there is an inequality

the Anaxagorian argument, we have shown in this work that continuous motion in the discontinuous, sensible universe is impossible. We need the unity, continuity, and simultaneity of the universe's discontinuous parts in order to have continuous motion. It follows that, far from being a principle of motion, discontinuous inequality and temporal order is an analytic principle of the interruption of continuous motion.

As we have argued, continuous motion is possible if, and only if, there is a limiting point, a cosmic singularity, to assign unity, continuity, constancy, and limit to the discontinuous, indefinitely accelerating, sensible universe. This cosmic singularity of maximum unity constitutes

between the opposite forces, if any of the two forces—for example, F—is greater than its opposite, the resistance R:

1. $$F \neq R \quad \rightarrow \quad F > R.$$

Given this principle of motion, it follows that if the opposite forces F and R are equal, then speed is zero and the body is at the state of rest. Galileo was the first modern physicist to deny this Aristotelian thesis and affirm the contrary, which in fact extends the Aristotelian thesis. According to Galileo, the physical body continues to move with a uniform motion if the above two forces are equivalent, such as:

2. $$F = R.$$

If we consider the equality $F = R$ as the synthetic product of opposite inequalities, we then obtain the following structure:

3. $$(F = R) = (F > R)(R > F),$$

where Aristotle's principle of motion—namely, the inequality $F > R$—is a particular case of the equivalence $F = R$ constituting the real, founding principle of uniform, continuous motion. As a particular case of continuous motion the analytic inequality $F > R$ constitutes a particular, incomplete or apparent principle of motion, which in reality interrupts continuous motion by virtue of its discontinuous nature. Its destructive effect on the continuity of motion is neutralized when we introduce at the same time the counter-inequality $R > F$, in order to obtain on the whole the equality of opposite forces that ensures the constancy and continuity of motion without destroying motion:

4. $$(F > R)(R > F) \rightarrow (F = R).$$

simultaneously the end of inequality and temporal order. We thus need the end of temporal order and the emergence of a constant rotational motion of maximum speed governed by the perfect equivalence principle in order to assign being, substance, and meaning to the time-conditioned sensible world and its rectilinear acceleration. In fact, a rectilinear acceleration without a constant limit to transform it into a higher rotational motion of maximum speed comprising the rectilinear acceleration as a particular case is an infinitely regressing acceleration, which is groundless and self-defeating.

The *end* of growth, therefore, is not *indefinite* growth but rather *maximum* growth. We have maximum growth when we experience through our infinite, universal sense the real 1—the real, physical body from all sides—and when we experience its supreme good—the perfect equivalence principle that unites (via its cosmic singularity) all its parts.[10]

With these points in place, it is easy to see that if our brain possesses the faculty of infinite, universal sensibility, then any indefinite extension of our finite, particular sensibility through artificial means is useless. It follows that the *end* of technology is not the indefinite growth of artifacts indefinitely expanding our finite, particular sensibility. Rather,

---

10. Given the incompatibility between temporal order and continuous communication, we are led to conclude that Aristotle's principle of noncommunication of different things belonging to different categories or classes is applied uniquely to the inner observable part of the physical universe governed by the analytic principle of inequality and temporal order. Per contra, Plato's principle of continuous communication of different categories is applied to the real, physical universe itself with a cosmic singularity and governed by the perfect equivalence principle. The physical universe composed of parts that immediately communicate via the cosmic singularity constitutes a universal community of infinite wholes. Therefore, a universal community of things is manifest as soon as we perceive the infinite succession of things from outside—from the cosmic singularity—as cosmic unity and simultaneity.

the *end* of technology is the attainment of the faculty of infinite, universal sensibility. This faculty will enable us to experience directly and completely the real 1—namely, the divine state of *first being* and its supreme good, the perfect equivalence principle—independently of extrinsic, artificial means.

The Geometric Synopsis of the Infinite in Act

The Greek philosopher Zeno discovered that if we add successively, one by one, ab's infinite number of parts of the general form $(1/2^n)$, we will never arrive at b, the maximum limit and total sum of parts. In fact, we will always be as far off from b as is the starting point a. He concluded that motion from the starting point a called 0 to the final point b called 1 or—to put it another way—from the unlimited series of parts a_n, taken as the starting point $a = 0$ to the end-point $b = 1$ is impossible.

Because no matter how many parts we add successively to the unlimited series of parts a_n we will never reach the limit and total sum 1, the solution is not computational or quantitative. Rather, it is noncomputational and qualitative. Thus, what we need is a principle that will connect the unlimited a_n with something entirely different from itself, which is the limit 1, and therefore will allow by this connection the continuous and immediate motion from a_n to 1.

This principle is the *synthetic principle of equivalence*. In its general form, this principle stipulates the equality of unequal things; in its particular form, it stipulates the equality of the infinite and the finite, of the *unlimited* series a_n with the *limit* 1: $a_n = 1$.

If continuous motion is a real feature of the physical universe, then the real, physical universe is continuous composed of equal parts, the complex, infinite wholes, that verify synthetic principles of being: for example, the finite–infinite equivalence principle, which states that anything finite is infinite and anything infinite is finite.

Grounded in the finite–infinite equivalence principle, we define the maximum limit 1 as anything finite, which is simultaneously infinite according to extension and division. It follows that the maximum limit 1 is the synthetic product and constant ratio of infinite and zero magnitudes: $1 = \infty \times 0$, $1 = 0/\infty$, where ∞ is the infinite magnitude according to extension and 0 is the infinite magnitude according to division. Both synthetic definitions show that we need at least three magnitudes to compute exactly the maximum limit 1 whose number is the real 1.

The real 1 designates the real, physical body of maximum unit radius $ab = 1 = \infty \times 0$, rotating continuously at the maximum speed of light $c = 1 = 0 \times \infty$. The physical body, whose substance is the infinite whole, is with respect to extent the limit and total sum of an unlimited series a_n of extended parts of increasing and decreasing magnitude and with respect to speed the constant and maximum limit of an infinite series a_n of accelerated parts of increasing speed. In this sense, the physical body is both the greatest and the smallest, the highest body and the fastest, occurring in the ultimate past or the ultimate future, and which we call the *first body* governed by the perfect equivalence principle. This synthetic principle, which stipulates the equality of all parts of the first body, is the body's highest intelligence, its highest truth, and supreme good.

We divide the real, physical body of unit radius 1 into the physical universe (object), constituting the maximum limit 1 of our accelerating, sensible universe a_n, and into the universal brain (subject) constituting the maximum limit of our accelerating individual brain a_n. Thus, the real, physical body is simultaneously a perceived object—that is, the physical universe—and a perceiving subject—that is, the universal brain. In total, the physical body is a self-perceiving body; it *is* and has a complete perception of its *being*. The principle that grounds self-perception is the principle of self-order or self-causality applied to the physical whole.

According to self-causality, anything that is a continuous, physical whole is simultaneously an antecedent cause and a consequent effect; it is an antecedent perceiving subject and a consequent perceived object or predicate. Self-causality is a variant of the synthetic principle of perfect equivalence.

Let us represent the unlimited series of parts a_n by the open, unlimited, Euclidean (or hyperbolic) plane and the limit and total sum 1 by a closed, limiting sphere (greatest sphere) of center a and maximum unit radius $ab = 1$. The geometric one-dimensional figure of the real, physical universe is the intersection of the limiting sphere and the Euclidean plane passing through the center a of the limiting sphere. We call this intersection the limiting circle (greatest circle) of maximum unit radius $ab = 1$, where a is the fixed center and b is the limiting circumference rotating at the maximum speed of light and vibrating a maximum number of cycles per second.

Seen from inside, the limiting-circumference b is a Euclidean (or hyperbolic) line of infinite radius identified to the inner flat (or concave) part of the limiting-circumference b and containing the unlimited series of consecutive and isolated parts a_n. Seen from outside, the limiting-circumference b is a spherical limiting-point b of zero radius and infinite curvature, which we call *space-time singularity* or *cosmic singularity*, assigning unity and limit to the unlimited series of isolated parts. Because this cosmic singularity has a zero radius and an infinite curvature, it occurs in dimension zero and has infinite energy, infinite density, and infinite temperature. The synthetic product of infinite and zero magnitudes determines a complex, indeterminate limiting-circumference b, an open-closed circumference, of maximum unit radius $ab = 1$ defined as the complex product and constant ratio of infinite and zero magnitudes numbered by the real 1: $ab = 1 = \infty \times 0$, $ab = 1 = 0/\infty$.

If unity is proportional to the curvature of the physical universe, then the cosmic singularity of infinite curvature infinitely unifies the parts of the unlimited plane regardless of their difference and distance. We consider, then, the unifying cosmic singularity as the material and geometric realization of the perfect equivalence principle, which stipulates the unity of all diverse parts of the physical universe and hence their immediate and universal communication.

Different observers O and O' located at different places in the physical universe possess different levels of energy and speed and have different faculties of sensibility for perceiving the real, physical universe.

The observer O is a particular observer situated at the center a of the physical universe, the region of middle scales. In this intermediate region, O's individual brain is composed of individual cells at the state of rest and having a low-energy content. Because of their resting state and low-energy content, the individual cells (retinal cells) detect only a finite portion of physical light, which we call sensible (optical) light of definite wavelength λ traveling through space-time uniquely along a diagonal axis, in a unique (forward) sense, at the unique, finite speed c, which we take as equal to 1. Finite, sensible light reveals to O's individual brain the inner, past part of the physical universe composed of consecutive, isolated parts, which form the sensible universe. The mental faculty of perceiving the physical whole uniquely from inside and hence partially and successively in conformity with the analytic principle of inequality and temporal order we call *finite, analytic, particular sensibility*. This incomplete mental faculty belongs to O's individual mind whose organ is the low-energy, individual brain.

The inner, sensible part of the physical universe appears as an unlimited, Euclidean (or hyperbolic) plane a_n without a limiting-point b (a cosmic singularity) to limit it and unite its unequal parts. The Euclidean plane a_n, without a limiting-point b, is equivalent to a sensible universe, which indefinitely accelerates and extends to the limiting-point b occurring either before or after the Euclidean, sensible universe a_n but never simultaneously with the sensible universe a_n.

The sensible universe a_n is composed of unequal, consecutive, isolated parts (the individuals) deprived of unity and continuous motion and verifying analytic principles of being, such as the principle of inequality and temporal order. Thus, no matter how much the sensible universe a_n (or the series of sensible universes a_n) approaches its respective limiting-point b, it is always as far off from the limiting-point b as is the first member a_0 of the series at the starting point and center a. Continuous motion from a to b is impossible insofar as the mental faculty of finite, particular sensibility perceives the parts a and b of the physical universe, where a designates a_n and b designates 1, discontinuously and successively: $a \neq b \rightarrow a < b$. It follows that if there is continuous motion from a to b, it is because in reality the physical universe has a cosmic singularity unifying its discontinuous, unequal parts in conformity with the synthetic principle of equivalence, which stipulates the equality of unequal parts: $(a = b)\,(a < b)$.

It is not sufficient to provide a geometric solution to the problem of motion by assuming the existence of a cosmic singularity unifying the unequal parts of the physical universe; we must also experience this *cosmic unity* if we want to realize effectively the continuous passage from $a = a_n$ to $b = 1$, and if we want to experience effectively the real 1, the end-point of a_n's accelerated motion. To have such an experience, we must assume the existence of an ideal, cognitive faculty. This cognitive faculty is the infinite, universal sensibility belonging to the universal brain and having the power to perceive the physical universe spherically, from all senses and sides, from inside as an indefinitely extending Euclidean (or hyperbolic) plane and from outside as an extensionless, limiting point, as a cosmic singularity unifying all Euclidean parts and transforming the indefinitely extending part a_n into an infinitely extended whole 1.

We obtain this highest faculty of infinite, universal sensibility by elevating our low-energy individual brain and its mental faculty of finite, particular sensibility to the level of highest unity, the cosmic singularity level, which is the outer, infinitely curved, convex part of the physical universe's limiting-circumference b. We therefore call *universal brain* the low-energy individual brain elevated to the level of highest unity, energy, and speed in order to possess the highest mental faculty of infinite, universal sensibility. We call *universal observer* O' the observer with a universal mind whose organ is the universal brain and who is situated on the universe's outer limiting-circumference b.

Because unity is proportional to the body's curvature and energy, it is sufficient to elevate maximally the energy content of the brain's individual cells in order to obtain a universal brain composed of universal cells that have a maximum energy content. These highly energetic universal cells emit or receive the entire physical light traveling simultaneously through space-time and beyond space-time (via the cosmic singularity) at all senses and frequencies, and revealing at once the entire physical universe, from inside and outside, its infinite past and its infinite future.

The elevation of the energy content of the brain's cells can be undertaken through the maximum contraction of their size or through the maximum acceleration of their speed. Both actions must attain the cosmic singularity level of infinite curvature and energy located at dimension zero. We call *organic singularities* the brain's universal cells that have touched the cosmic singularity of infinite energy and assign to the brain the mental faculty of infinite, universal sensibility, and the physical power of permanent life and universal communication.

By *permanent life* we mean the natural power to exist anywhere, at any time, under any condition. By *universal communication* we mean the natural power to communicate immediately and continuously, via the cosmic singularity, with any part of the physical universe regardless of its difference, distance, and direction. In this way, the universal brain has the natural power to perceive and to travel through the universe's unlimited series of parts a_n at once, and thereby to experience the indefinitely varying part a_n as a constant, infinite whole 1 composed of equal, simultaneous parts in continuous, immediate communication.

But how is it possible to elevate maximally the energy content of the individual brain and thus to attain the cosmic singularity level of infinite curvature and highest unity without being crushed into nothingness? The elevation of the brain to the cosmic singularity level situated on the universe's outer limiting-circumference b does not not crush the brain because already and since eternity the brain, analogous to the physical universe, is the greatest complex body occupying at all times both places, the inner center a where the curvature of the universe is zero and the outer limiting-circumference b where the curvature of the universe is infinite. In this way, the brain is not smashed by the infinite curvature, as this infinite curvature is neutralized by the universe's zero curvature. Ultimately, the unit curvature of the physical universe emerges at any point of the universe as the constant and balanced ratio of zero and infinite curvatures, which mutually neutralized neither crush nor dilute the brain.

The simultaneous occupation of contrary (opposite and equal) places by the universal brain is possible because the inner center a and the outer limiting-circumference b are equal and simultaneous sources of the greatest circle. Indeed, we need both parts a and b in order to describe the greatest circle of greatest unit radius $ab = 1$ representing one-dimensionally the greatest body of the physical universe and its universal brain. Because the opposite parts a and b are (according to the center–circumference equivalence principle, which governs the greatest circle) equal, we can interchange their properties and have: $ab = ba$, where the cold center a of zero curvature is simultaneously on the hot limiting-circumference b of infinite curvature and the hot limiting-circumference b of infinite curvature is simultaneously at the cold center a of zero curvature (see fig. 4.7, p. 159).

Thus, relative to the particular observer O at the center a, the brain occurs at the cold center a of zero curvature and possesses the following fundamental properties:

- It is a low-energy individual brain at the state of rest or of indefinite acceleration according to extension and division.
- It is conditioned by asymmetric time and behaves in conformity with analytic principles of being—for example, in conformity with the principle of inequality and temporal order.
- It detects only a finite portion of physical light, revealing the physical universe uniquely from inside, as an unlimited series a_n of discontinuous, consecutive parts without the limiting-point 1 to assign unity, communication, and justice between the parts.
- It possesses the cognitive faculties of finite, particular sensibility and analytic understanding, regarded as the faculties of knowing the physical world partially from a unique side.

On the other hand, relative to the universal observer O′ on the limiting-circumference b, the same brain occurs on the hot limiting-circumference b of the physical whole and possesses the following properties:

- It is a high-energy universal brain permanently rotating at the maximum speed of light and permanently vibrating with a maximum frequency.
- It is timeless and behaves in conformity with synthetic principles of being—for example, in conformity with the synthetic principle of equivalence and zero-temporal order.
- It detects the entire physical light, revealing the physical universe from inside and outside, as an unlimited series a_n of continuous, simultaneous parts with the limiting-point 1 to assign unity, communication, and justice between the parts.

- It possesses the cognitive faculties of infinite, universal sensibility and synthetic reason, regarded as the faculties of knowing the physical universe spherically from all sides.

The brain is a universal being occurring everywhere, in all directions, at all times, under all conditions: at the cold center a of zero curvature as a resting individual or as an indefinitely accelerating and extending part a_n and on the hot limiting-circumference b of infinite curvature as an infinitely extended physical whole with a maximum unit radius ab equal to the real $1 = \infty \times 0$ and permanently rotating at the maximum speed of light $c = 1 = 0 \times \infty$ (see fig. 4.7, p. 159). Indeed, it is the immanent property of the synthetic, universal brain to be unaffected by any of these particular conditions, by the cold center a of zero curvature and the hot limiting-circumference b of infinite curvature, and hence to be an impartial (neutral) and unconditioned being—that is, free of all particular conditions. As an unconditioned being, the brain can realize the hot cosmic unity of infinite curvature *here* and *now* at the cold center a of zero curvature, where Euclidean parts are isolated, unequal, and unjust, without burning and crushing: without disintegrating its structure. This is not a miracle or magic; it is the very immanent nature of the greatest body, the physical universe and the universal brain of maximum unit radius $ab = 1 = a/b$ emerging as a constant proportion, equality and balance, between maximally different and distant parts, between the cold center a of zero curvature and the hot limiting-circumference b of infinite curvature.

To the particular observer O, acceleration may appear to be the *indefinite* extension of our finite, particular sensibility through artificial means. In reality, however, and relative to the universal observer O′, the meaning of acceleration and extension is the attainment of the highest, universal sensibility that will enable us to experience the real 1 independently of extrinsic, artificial, temporal, or causal means. Indeed, if we use external, artificial, temporal, or causal means for attaining the real 1, what we accomplish is merely the indefinite approach of the real

1, which remains permanently external, inaccessible, and greater than the series of artificial means. What we need, therefore, for the effective attainment of the real 1 is the activation of our proper infinite, universal sensibility naturally residing within us since eternity.

If the future of our accelerating sensible universe a_n is the effective realization of the real 1, then science, or, better, *scientific metaphysics*, must tell us how to activate our dormant infinite, universal sensibility in order to experience naturally, without constraint and effort, the state of highest being possessing its highest good, the perfect equivalence principle.

Glossary

Absolute. (1) The being in itself. (2) The constant. (3) The independent (or unconditioned). See also **substance.**

Accident (κατά συμβεβηκός). (1) The being relative to something else or as something else. (2) What appears to (or is dependent on) the particular sense. The *accident* is variable and conditioned. See also **relative.**

Actual infinite. See **infinite.**

Agnosticism. See **nihilism** (epistemological).

Analytic logic. The science of the principles that govern the object experienced as an individual (particular), having at one time a unique determination or contradictory determinations and at different times consecutive determinations. See also **principles of organization of the object.**

Analytic understanding. See **mind** (faculties of).

Analytic universal proposition. A determinate (categorical) proposition having the form *every x is either a or a'*. This Euclidean proposition either affirms or denies something of the object universally. The object is experienced as a simple individual having at one time a unique determination (part, sense) and verifying the analytic principles of contradiction and excluded third.

Appearance. (1) Neutrally, a presentation to an observer. (2) Metaphysically, a part of the object—a part of objective truth. It follows that its content of reality is partial (incomplete), relative, and observer-dependent.

Archimedes, axiom of. Given a numeric magnitude a of a given kind, no matter how small or how great a is, we can always find a numeric magnitude b that is either less or greater than a. According to the axiom of Archimedes, there are no minimum or maximum numeric magnitudes in the Euclidean world.

Artificial. That which is unnatural or nonspontaneous, and depends on an external cause for its existence and motion. The artificial is ephemeral; its fate is collapse.

Being. We consider the extended thing, body, and whole as metaphysical equivalents of the being. When the being is the aim of cognition, we call it the *object*.

Body. (1) That which has a complete reality being the sum total of all partial realities. The body is therefore a complete whole whose geometric figure is the limiting sphere. Because it is assumed that the terms *thing* and *being* have a complete or maximum reality, we consider them to be synonymous with the body. When the body is an object of cognition, we call it the object. The body or object can be the physical universe and anything in the physical universe. (2) If the body is that which is limited and defined by a surface (Aristotle), then the limiting surface of the spherical body defines the sum total of all its parts. It follows that the body's total sum of parts is proportional not to its volume but rather to its surface area.

Brain. We distinguish between the **individual brain** and the **universal brain.**

- The **individual brain** is the brain in the state of rest or indefinite acceleration (infinite regression), conditioned by time and occurring at the levels of low energy and low unity, where space-time has zero curvature. The individual brain perceives only an arbitrarily selected part of physical light, called sensible (observable) light, revealing only a part of the physical universe, called the sensible (observable) universe.

- The **universal brain** is the brain elevated to the level of highest energy and highest unity, where space-time has infinite curvature. At this highest level, the permanently and maximally rotating universal brain perceives the entire physical light, revealing at once the entire physical universe—its inside and outside, its past and future.

Circle (*ideal* or *real circle, greatest circle, limiting circle*). The limiting circle assimilated to the *limiting-circumference b* is formed by the intersection of the greatest sphere and the unlimited Euclidean plane passing through the limiting sphere's center *a*. The limiting circle of center *a* and maximum unit radius *ab* represents one-dimensionally the physical universe, which is the greatest magnitude according to extension and division. Seen from inside (the center *a*), the limiting circle has an infinite radius according to extension; seen from the outside the limiting circle has a zero radius according to division. The product of infinite and zero magnitudes gives the finite magnitude 1: $ab = \infty$

$\times\ 0 = 1$. In total, the limiting circle, of finite unit size, is an infinitely extended straight (or hyperbolic) line when seen from inside and an extensionless point, a cosmic singularity, when seen from outside.

The finite, dimensioned magnitude of the limiting circle's infinite radius is equal to the cosmological distance 10^{26} m, whereas the finite, dimensioned magnitude of the limiting circle's zero radius is equal to the quantum distance 10^{-34} m. Both magnitudes express the maximum distance according to extension and division that the limiting circle has relative to us (at the center a).

The limiting circle is the region where an indefinitely varying, finite quantity becomes a maximum quantity, a real, infinite whole 1 defined as the product or constant ratio of infinite and zero magnitudes: $1 = \infty \times 0$, $1 = 0/\infty$. Finally, it is the region where galaxies and brains attain the state of *first being* governed by the synthetic equivalence principle, moving permanently at the maximum speed of light and having in the case of brains the mental faculty of infinite, synthetic, universal sensibility.

Concave. A surface curving inward or away from an observer. Its negative curvature causes the divergence of things away from the eye at the center a. The inner surface of a sphere appears concave (or hyperbolic).

Contradiction. The impossible unity and coexistence of opposite things. Absurdity, falsity, paradox, illusion, and schizophrenia are different manifestations of the impossible unity of different (opposite) things. The impossible unity and coexistence of different things are hate and sorrow with respect to feeling; unintelligence and falsity with respect to comprehension; insanity and injustice with respect to action.

Contradiction (principle of). See **principles of organization of the object.**

Contradictories. Opposites whose unity and coexistence are impossible. Contradictories exist successively in time. There is nothing intermediate between contradictories to realize their unity by contact; thus their opposition is absolute and analytic. Motion between contradictories that are deprived of unity is impossible or incomplete.

Contraries. Opposites whose unity and coexistence are necessary. There is an intermediate thing between contraries that realizes their unity by contact; therefore, their opposition is relative (partial) and synthetic. All motion is between spatial contraries. Spatial contrariety is: (1) the maximum unity of maximally different things and (2) the immediate contact between maximally distant things.

The *first contraries* are the spatial opposites—*low* at the inner center *a* (here and now) and *high* on the outer limiting-circumference *b* of the physical universe (Aristotle); their product or ratio is unity: $ab = 1$, $a/b = 1$. In the infinitely small and infinitely great circles, the center *a* and the circumference *b* coincide (Nicolas of Cusa).

Convergence. (1) Analytic definition: The property of a series of numbers in which the difference between consecutive terms gradually decreases but always remains greater than zero. The sum of converging series approaches the limit as the number of terms tends to infinity. (2) Synthetic definition: The property of a series of numbers in which the difference between simultaneous terms is both greater than zero and zero. The sum of converging series reaches the limit as the number of terms reaches infinity.

Convex. A surface curving outward or toward an observer. Its positive curvature causes the convergence of things toward the eye at the center *a*. The outer surface of a sphere appears convex.

Cosmic singularity. A region where the curvature of space-time is infinite and the size of space-time is zero. Physicists call it space-time singularity; we call it cosmic singularity or limiting-point *b*. Because of its size zero and infinite curvature, the cosmic singularity has infinite density of matter, and is the source of infinite unity and infinite curving force (attractive gravity). The cosmic singularity of size zero can be regarded as the smallest distance between bodies, called unity, and as the smallest body, called nothing. In a general manner, a region where any finite quantity of a given kind takes an infinite magnitude according to extension and division.

- A **transcendent cosmic singularity** is *absolutely beyond* the sensible world and occurs either before the sensible world as its violent big bang or after the sensible world as its fiery big crunch in conformity with the analytic principle of temporal order. In both cases, transcendent cosmic singularity is a black hole emitting no light, or, to put it another way, light of zero speed (or infinite wavelength) and zero frequency. Geometrically, we identify it with the *simple* limiting-point *b*, which is uniquely intellectual and intelligible.

Glossary

- A **transcendental cosmic singularity** is both *beyond* and *within* the sensible world, occurring everywhere at all times in conformity with the synthetic principle of zero temporal order. In fact, it constitutes the outer part of the inner, sensible world and together with the sensible world forms the real, physical whole, which is a luminous object emitting light of maximum speed (or maximum wavelength) and maximum frequency. Geometrically, we can identify it with the *complex* limiting-point b, which is both intellectual and sensuous, intelligible and sensible.

Curvature. In one-dimensional space, the curvature K of the circle is in inverse proportion to that of its radius r: $K = 1/r$. If the radius r is infinite, then the curvature of the circle is zero and the circle is a straight line. If the radius r is zero, then the curvature of the circle is infinite and the circle is a spherical point.

Diagonal line. This divides the whole 1 represented by the square's diagonal line u into opposite and equal parts a and a'—for example, into horizontal and vertical parts—and at the same time unites the opposite parts. In this sense, the diagonal line is both a dividing line and a proportionality line. **Circular line** has the same properties as the diagonal line; it divides the whole 1 represented by the circular line of curvature 1 into inside and outside, or its unit curvature into zero and infinite curvatures, and at the same time is the synthetic unity and contact of inside and outside, of zero and infinite curvatures. We can similarly define the diagonal or circular line as the set of all things that are complex—namely, both a and a'— and indeterminate—namely, neither a nor a'.

Dogmatic metaphysics. See **metaphysics.**

Epistemological illusion. See **illusion.**

Epistemological nihilism. See **nihilism.**

Error. A falsity that is taken as truth.

Euclidean geometry. The geometry of the sensible (observable) universe composed of simple individuals having at one time a unique determination or contradictory determinations, at different times consecutive determinations, ruled by analytic principles of organization. In the Euclidean universe, the center a is everywhere and the circumference b is nowhere. The radius of the Euclidean universe is unlimited without limit. Similarly, the Euclidean, sensible universe is a finite variable increasing (or decreasing) indefinitely in time. Because in the Euclidean plane different points $a \neq b$ are unequal, and determinate and verify analytic principles of organization, the Euclidean plane is nonuniform or partially (locally) uniform. For example, it is uniform in its space dimension but not in its time dimension.

Euclidean world (space-time). (1) The impossibility to inverse the body's sense a as it moves away from the starting point. (2) The supposition that as we progress in the world, we are constrained *either* to recede from the starting point, and in this case we conclude that the world is open, flat, or hyperbolic, *or* to return to the starting point, and in this case we conclude that the world is closed and spherical.

Finitism. The impossibility of assigning infinite extension and infinite division to the finite body equal to itself and one; hence, the impossibility of assigning infinite and zero magnitudes to any of its sensible (observable) quantities taken as a unit—that is, as equal to 1.

First (πρώτη, primary, original). That which is independent of, and ontologically and chronologically prior to, our particular perception and which we identify with the physical universe.

- By **first being** (ideal, real, physical being) we mean the being of beings (*ens entium*)—the supreme being—which is the physical universe. The first being occurs ontologically and chronologically prior to us and is situated on the outer limiting-circumference b of the physical universe, where the cosmic singularity emerges and space-time begins or ends. Located in the ultimate past, the first being is the *first cause* of a series of parts; located in the ultimate future, the first being is the *final cause* of a series of parts. See also **supreme being.**

- The **first body** (ideal, real physical body) is the body of bodies—the highest body of the first being whose limiting boundary is the physical universe's limiting-circumference b.

- **First motion** (ideal and real motion). The motion of the first body changing place within an instant (immediately) at the maximum speed of light.

- The **first principle** is the synthetic equivalence principle, which is the principle of supreme equality (unity), intelligence, truth, justice, harmony, and good. It governs the first body of the first being, which is assimilated to the physical universe in continuous rotation at the maximum speed of light.

- The **first quantity** is any kind of maximum quantity that has the magnitude 1 and is the product or constant ratio of infinite and zero magnitudes: $1 = \infty \times 0$, $1 = 0/\infty$. We assign to the first quantity the real number 1. Because the real 1 is the product of infinite and zero magnitudes, where the infinite is the limit of an infinite sequence of increasing magnitudes and zero is the limit of an infinite sequence of decreasing magnitudes, the real 1 is ultimately an infinite totality of increasing and decreasing magnitudes. By comprising an infinite totality of increasing and decreasing magnitudes at once, the first quantity of a given kind varies according to extension and division independently of time.

Force F_u. The original or *first force*, which is also called the ideal, real force, resides in the complex, indeterminate mass of an extended material body. The original force F_u is the synthetic unity and constant ratio of attractive gravity F, which is a contracting or curving force unifying things, and of repulsive gravity R, which is an expanding or stretching force separating things: $F_u = F/R = 1$. On the whole, force F_u has both attractive and repulsive, curving and stretching, centric and eccentric properties. The nature of F_u may be gravitational or electromagnetic, or neither gravitational nor electromagnetic. See also **mass**.

Gnosticism. The epistemological doctrine asserting that knowledge (empirical and intellectual) of the universe and communication of this knowledge in the universe are necessary and certain. Using the synthetic principle of reflexive order, gnosticism asserts that the universe, along with the sum total of all its knowable objects, can be itself an object of knowledge (this is also a basic claim of scientific and mathematical metaphysics). Gnosticism's opposite, agnosticism (skepticism), uses the analytic principle of irreflexive order or temporal order and asserts that the universe, being the sum total of all its knowable objects, is not itself an object of knowledge (this is also a basic claim of dogmatic metaphysics). See its opposite **nihilism** (epistemological).

Greatest Circle. See **circle**.

Ideal motion (instantaneous motion at a distance). See **motion**.

Illusion. Something taken as something else (general definition). From the metaphysical point of view, we have two kinds of illusion:

- **Perceptual illusion.** The perception of the complete whole as being uniquely an incomplete part. For example, the perception of the physical object thought of as a complex, infinite whole free of time *as if* it were uniquely a simple, finite part, an individual, conditioned by time. This perceptual illusion is the consequence of the low-energy individual brain being at the state of rest and arbitrarily perceiving only a finite part of light, revealing only a finite part of the entire physical object. The finite, sensible part of light we call *sensible light*, whereas the finite, sensible part of the physical object we call *sensible object*.

- **Ontological** and **epistemological illusion.** The understanding of the incomplete part *as if* it were a complete whole. For example, the understanding of the time-conditioned, sensible part of the physical object *as if* it were the real, physical object itself, and the understanding of the analytic principles of our particular perception prescribed to the time-conditioned, sensible part *as if* they were the objective principles of the real, physical object.

    However, if the complete whole and the incomplete part are assumed to be equally true and real, then to perceive the complete whole as an incomplete part—something as something else —is not a perceptual illusion but rather an alternative way of perceiving the real, physical object, which as a complex, infinite whole is both a complete whole and an incomplete part and thus has alternative ways of being.

Impossible. That which is deprived of power and does not admit contraries.

Individual (*particular*). (1) Anything that admits at one time a unique determination and contradictory determinations and at different times successive determinations. (2) Anything equal to itself and one and numbered by the finite number one. According to Aristotle, things are individuals obeying analytic principles of organization. See **principles of organization of the object, being, thought.**

Infinite (ἄπειρον). Aristotle divided the infinite into the infinite as complete whole (infinite whole, actual infinite, absolute infinite) and into the infinite as incomplete part (potential infinite, relative infinite):

- **Infinite whole** or **actual infinite** (ἄπειρον ἐν ενεργεία). (1) The infinite multiplicity of things unified into one thing in order to constitute the infinite whole (general definition). (2) The limit and total sum of an unlimited series of parts. (3) The potential infinite (indefinite) actualized, and completed at the limit. (4) The infinite as maximum, which has nothing external to itself and beyond itself—the proper infinite. (5) The spatial infinite according to extension and division, which is free of time (of succession-comparison-computation) and the laws of arithmetic. (6) The complex (synthetic) infinite, which at one time has opposite and equal or coexisting determinations (the contraries)—as for example, the unlimited and the limited. (7) The neutral (impartial) infinite, which is beyond and between the opposite and equal determinations for example, the unlimited and the limited), and which we call the indeterminate and the intermediate (τό μέσον). (8) The internally unbounded infinite (Anaximander, sixth century BCE) without internal distinctions between its opposite determinations, which coincide on the circle. (9) The limiting circle (greatest circle), which is the intersection of the limiting sphere and the Euclidean plane passing through the center of the limiting sphere. The limiting sphere closes, delimits, and unites the open, unlimited Euclidean plane into an infinite whole 1.

  The infinite whole according to extension is denoted by ∞ and the infinite whole according to division is denoted by $1/\infty = 0$. Ultimately, the infinite whole equal to itself and one (the infinite one) is the *synthetic product* or *constant* and *balanced ratio* of infinite and zero magnitudes to which we assign the real number 1: $1 = \infty \times 0$, $1 = \infty/0$. See also **first quantity**.

  The founding principle of the infinite whole is the synthetic principle of equivalence, particularly the finite–infinite equivalence principle, which stipulates the equality of the finite and the infinite—that everything unlimited is limited and everything limited is unlimited. Other founding principles of the infinite whole are that of reflexive order (self-order) and that of self-causality. Finally, we use our spatial faculties of synthetic reason and transcendental imagination for intellectually knowing the infinite whole.

  According to Aristotle, the infinite whole or actual infinite is an absurdity because it denies the analytic principle of contradiction, affirming that nothing unlimited is limited. However, for Plato and Pythagoras, as well as for the majority of the Greek Ionian philosophers (Melissus, Anaximander, Anaxagoras, Leucippus, Democritus), the infinite whole (that which is both unlimited and limited) is the very immanent nature of the real, physical body.

- **Potential infinite** (ἄπειρον ἐν δυνάμει). The infinite, which always has something external to itself and beyond itself (Aristotle) and verifies the *axiom of Archimedes*. The potential infinite is an improper infinite; in reality, it is a finite part indefinitely varying in time and obeying the laws of arithmetic. The one-dimensional geometric figure of the potential infinite is the incomplete straight line to which we can always add something (Aristotle). The straight line contains an infinite multiplicity of parts without a limiting point to delimit and unite it into an infinite whole. An infinite without limit—namely, the *indefinite*—is a simple infinite (analytic infinite) that verifies analytic principles of organization, particularly the analytic principle of reflexive identity, which stipulates the equality of the infinite with itself, the infinite. Because the indefinite variation (infinite regression) of the finite part destroys its motion, the indefinite or potential infinite is also an evil infinite.

  We use our sensuous intuition of time and our finite faculties of analytic understanding and particular sensibility to know the potential infinite intellectually and sensuously.

Infinitism. The assumption that the finite body equal to itself and one is infinitely extended and infinitely divided—that is, an infinite whole having simultaneously infinite and zero magnitudes.

Irrealism. See **nihilism** (ontological).

Light. We divide light into **real, physical light** (ideal light, first light), carrying information about the entire physical universe, its inside and outside, its past and future; and *sensible* light, carrying information about only the inner and past part of the physical universe. Physical light of definite wavelength λ travels through space-time in all directions, at all speeds c, which are equal to less than and greater than 1, where 1 is equal to 3×10^8 m/s: $(c = 1) = (c < 1)(1 < c)$. We can also say that physical light, which is a real, infinite whole 1, is spherically propagated at a maximum speed c equal to the real 1 and defined as the product of zero and infinite magnitudes: $c = 1 = 0 \times \infty$. On the other hand, sensible light, which is the finite, observable part of physical light of definite wavelength λ, travels space–time in a unique direction (Euclidean propagation) at the unique, sensible speed $c = 3 \times 10^8$ m/s = 1.

Light-year . The distance traveled by sensible light in one year (or 3×10^7 seconds). The speed of sensible light is 3×10^8 m/s. This distance is: 3×10^8 m/s $\times$ 3×10^7 s $= 10^{16}$ m

Limiting circle. See **circle.**

Luminosity (intrinsic). The rate of radiation of light into space by a star or other object.

Mass. The amount of matter contained in a permanently moving body. This mass m is divided into gravitational mass m_g originating F, the force of gravity (attractive gravity, active force, contracting-unifying force, positive curving force causing the convergence of parts toward the body's center) and inertial mass m_i originating R, the opposite and equal force of antigravity (repulsive gravity, resistance, stretching-separating force, negative curving force causing the divergence of parts away from the body's center). In a three-dimensional body, F is inversely proportional to the square of distance between the parts and the body's center; its inverse $R = 1/F$ is proportional to the square of distance between the parts and the body's center. The constant ratio $F/R = 1$ determines the equivalence $F = R$ of the opposite forces in which the balance and permanence of the moving body is grounded (see also **force F_u**). If any of the contrary forces is greater than the other force, then the balance and permanence of the moving body is broken, eventually causing its hot or cold death. Indeed, if $F > R$, then the contracting body implodes into a hot point without extension. If $R > F$, then the exploding body dilutes into a cold line without unity and energy. A balanced and permanently moving body is beyond explosion and implosion, expansion and contraction; it moves on a limiting circle continuously and immediately through and beyond Euclidean space-time.

Mathematical metaphysics. See **metaphysics.**

Maximum. (1) Any magnitude simultaneously admitting finite, infinite, and zero magnitudes (the trinitarian nature of the maximum magnitude). (2) The finite magnitude, which is simultaneously, and in conformity with the finite–infinite equivalence principle, an infinite magnitude according to extension and division. Conversely, it is the infinite magnitude according to extension and division, which is simultaneously a finite magnitude. A maximum finite magnitude equal to an infinite magnitude is the Hubble constant 10^{26} m $= \infty$, whereas a maximum finite magnitude equal to zero magnitude is the quantum constant 10^{-34} m $= 0$. (3) The real 1 defined as the synthetic product or constant and balanced ratio of infinite and zero magnitudes: $1 = \infty \times 0$, $1 = \infty/0$.

Because a maximum magnitude is timeless, it is beyond the laws of arithmetic. For example, if we multiply a maximum finite magnitude a by itself an infinite number of times, we obtain an infinite magnitude a^∞, which is both greater than a and equal to a: $(a^\infty > a)(a^\infty = a)$. The maximum magnitude is both variable and constant.

Metaphysics. (1) The science of the constitutive principle of the real, physical being equally called first principle (see also **first**). The physical being can be the physical universe and anything in the physical universe. (2) The science of the real, physical being defined as the sum total of an infinite series of parts. (3) The science of the physical being numbered by the real, infinite whole 1 and governed by the synthetic equivalence principle. We divide metaphysics into:

- **Dogmatic metaphysics.** If the sum total of all sensible parts is not itself a sensible part, then metaphysics and the series of sensible parts are dogmatic (Kant) and the nature of the physical whole is simple verifying analytic principles of organization.

- **Mathematical** or **scientific metaphysics.** If the sum total of all sensible parts is itself a sensible part, then metaphysics and the series of sensible parts are mathematical (Kant) and the nature of the physical whole is complex, therefore verifying synthetic principles of organization. By constituting, through the faculty of *universal sensibility*, the intelligible, physical whole a possible object of sense—a sensible thing—metaphysics becomes an observational science, a **scientific metaphysics.**

Mind (faculties of). We divide the faculties of the mind into four kinds:

- **Particular sensibility.** General definition: the faculty of sensing the private and particular (individual). (1) The faculty of perceiving through sensible light only an arbitrarily selected part of the physical object—namely, its inside and past—determining the sensible object composed of simple, consecutive, isolated parts. (2) The faculty of perceiving the particular properties of different things.

- **Analytic understanding.** (1) The faculty of thinking of the private and particular (individual) properties of things governed by analytic principles of organization. (2) The faculty of thinking of the infinite multiplicity of the physical object successively without unity and limit, called *arithmetic continuum* or *aggregate*. (3) The faculty of thinking of the parts of the physical object as simple, determinate things separated through categories and analytic principles, as, for example, the principle of inequality and temporal order.

- **Synthetic reason** (synonymous with the Greek λόγος or νούς). (1) The faculty of thinking of the common and universal properties of things governed by synthetic principles of organization. (2) The faculty of thinking of the infinite multiplicity of the physical object simultaneously, with unity and limit; the unity of the infinite multiplicity constitutes a *geometric continuum* and a *universal community*. (3) The faculty of thinking of the parts of the physical object as complex, indeterminate things maximally unified through ideas, limiting points (cosmic singularities), and synthetic principles, as, for example, the equivalence principle.

  A correlative faculty of synthetic reason is **transcendental imagination**. (1) The faculty of intuiting formally through intellectual images the physical object as different from what it appears to our particular sensibility or to our resting individual brain. (2) The faculty of intuiting the physical universe formally (intellectually) as a complex, infinite whole composed of complex, infinite wholes.

- **Universal sensibility.** General definition: The faculty of sensing the common and universal—that is, the cosmic singularity unifying all things regardless of their difference and distance. (1) The faculty of perceiving the entire physical light (*first light*) propagated at all speeds and frequencies and revealing the entire physical universe—its inside and outside, its past and future. (2) The faculty of perceiving the infinite multiplicity of the world simultaneously with unity and limit, constituting thereby a universal community free of conflict.

Motion. We divide motion into two kinds:

- **Instantaneous motion at a distance** (*first motion, ideal and real motion*). (1) The change of place within an instant at maximum speed (general definition). The speed of ideal motion is the maximum speed of light *c*, regarded as 1 and defined as the product or ratio of infinite and zero speeds: $c = 1 = \infty \times 0$, $c = 1 = 0/\infty$. By *infinite speed* we mean the speed that the moving body has when it travels one unit of space in zero time or when it travels infinite space in one unit of time. By *zero speed* we mean the speed that the moving body has when it travels one unit of space in infinite time or when it travels zero space in one unit of time. (2) The power to realize an infinite effect with one action or the power to realize one effect with zero action (effortless action); these are different aspects of the power to impart maximum motion without being moved, which constitutes the divine property of the Aristotelian *unmoved mover*. This last definition corresponds to Aristotle's term of immobile action (ενέργεια ακινησίας).

  Ideal motion is spontaneous (autonomous motion free of external force or cause), circular (reversible), permanent, continuous, immediate, uniform. The founding principle of continuous, ideal motion is the synthetic equivalence principle; the mechanism for communicating instantaneous action at a distance is the cosmic singularity.

- **Non-instantaneous motion at a distance** (*incomplete motion*). The change of place at different instants (successively) at the indefinitely accelerating finite speed. Because the founding principle of incomplete motion is the analytic principle of temporal order, it needs an external force (cause) for its existence. Incomplete motion is heteronomous (dependent on an external force or cause), rectilinear (irreversible), delayed (nonimmediate), discontinuous, and nonuniform.

Negation. An operation of change consisting of inversing the sense of a thing by a rotation of 180 degrees.

Nihilism. The doctrine of not-being—of nothing. We divide nihilism into:

• **Ontological nihilism** = irrealism. The doctrine asserting that nothing exists. Even if everything—that is, the universe—did exist, it could not *continue* to exist. Because ontological nihilism asserts the impossible continuity of the universe's existence, it necessarily predicts its violent collapse.

- **Epistemological nihilism** = agnosticism (skepticism). The doctrine asserting that the universe exists but that it is impossible to know its nature, which is assumed to be simple and determinate (Kantian skepticism). On the other hand, the ancient Greek philosopher Gorgias (fourth century BCE) asserted that (1) nothing exists; (2) even if everything—that is, the universe—did exist, it could not be known; (3) even if it were known, this knowledge could not be communicated.

  In both cases, nihilism uses analytic principles of organization for ontologically predicting the impossible existence of the universe and the impossible continuity of the existence of the universe, and epistemologically concluding the impossible knowledge of the universe, and the impossible communication of this knowledge within the universe (which is anological to Aristotle's principle of non-communicability between different things belonging to different categories).

Nous (νοὐς). See **mind,** faculties of.

Omnipotence. (1) The power to do everything (or anything) at once—for example, positing and denying a: $a = (a = a')$. (2) The synthetic power to admit contraries.

Omnipresence. The power to be present everywhere (or anywhere) in space-time by passing through the cosmic singularity occurring beyond space-time.

Omniscience. The power to know intellectually and sensuously everything (the universe) at any distance, in any direction, in any dimension, its inside and outside, its past and future, by emitting or receiving the entire physical light (*first light*) propagated at all speeds and frequencies.

The three divine properties of omnipotence, omnipresence, and omniscience characterize the *first being,* which is the physical universe, and presuppose the complete uniformity of the physical universe actualized by its unifying cosmic singularity.

One. We divide *one* into the finite whole 1, which is a number computing the object without parts, and into the real, infinite whole 1, which is both a number computing the object having infinitely many parts and an equivalence principle uniting the object's different parts. See **real 1.**

The finite whole 1 verifies analytic principles of being—for example, the principle of reflexive identity (intensive identity), according to which 1 is identical to itself: $1 = (1 = 1)$. In this sense, the finite whole 1 is an *individual constant* deprived of spatial contrariety and motion; it is thus an organic or sensible number computing the object (and any of its unit quantities) at rest. The real, infinite whole 1, on the other hand, verifies synthetic principles of being—for example, the synthetic principle of equivalence (extensive identity), according to which 1 is identical to something else: namely, to the infinite according to extension and division: $1 = \infty \times 0$. The real 1 is a *universal constant* receptive of spatial contrariety and motion; it is thus an ideal, physical number computing the object (and any of its unit quantities) at rest and at maximum motion.

Ontological illusion. See **illusion.**

Ontology. We divide ontology into two parts.

- **Analytic ontology.** The theory that everything is a simple individual, a finite whole 1 admitting at one time a unique determination (sense), and at successive times consecutive determinations, governed by analytic principles of organization.

- **Synthetic ontology.** The theory that everything is a complex universe, a real, infinite whole 1 admitting at one time contrary determinations (senses) and governed by synthetic principles of organization.

Organic singularity. A maximally synthetic state of being where an organic unit, the cell, has the infinite properties of the cosmic singularity without being disintegrated—that is, burned or crushed by its infinite curvature. In fact, the cell's infinite curvature is neutralized by its zero curvature producing thereby a unit curvature $K = 1$ that we define as the constant and balanced ratio of infinite and zero curvatures: $K = 1 = \infty/0$.

It is the state where our finite brain has infinite properties—for example, the power of infinite, universal perception that detects the entire physical light, its maximum speed (or maximum wavelength) and maximum frequency, revealing the entire physical object—the sum total of its parts.

Part. The incomplete thing that lacks everything (the universe). The incomplete part is simple and constant at one time, variable at different times, accidental, observer-dependent, and time-conditioned. The geometric figure of the incomplete part is the unlimited, Euclidean plane without a limiting sphere (cosmic singularity) to contain it and unify its parts. See also **sum**.

Particular sensibility. See **mind** (faculties of).

Perceptual illusion. See **illusion**.

Perfect (*ideal, complete, real*). See **whole**.

Permanent life. Continuous existence anywhere in space-time, in any direction, at all times, and under any condition.

Physical light. See **light**.

Physical universe. See **universe**.

Point (simple point). (1) That which has no parts—the individual. (2) A sphere that has zero radius and infinite curvature (see **singularity**). (3) That which has a position and a zero or finite extension.

Potential infinite. See **infinite**.

Principles of organization of the object (being and thought). We divide the principles of organization into analytic and synthetic principles. - The four analytic principles belonging to the empirical faculty of analytic understanding, which is the faculty of thinking of the object as a simple individual, are the following:

- **The principle of reflexive identity** (self-identity)—that everything a is equal to itself: $a = a$; that there is a sufficient reason why a is a and not something else not-a (Leibniz's principle of sufficient reason); that everything equal to itself is the finite number 1 (Aristotle): $(a = a) = 1$.

- **The principle of contradiction.** (1) That nothing is both something a and something else a' (not-a) called b: $(ab)'$; (2) that there is no equality and no communication between different things a and b belonging to different categories (Aristotle's principle of noncommunicability): $(a = b)'$; (3) that nothing a is unequal to itself: $(a < a)'$.

  If everything is simultaneously something else, and if everything is unequal to itself, we have then a contradiction, a paradox, an illusion, or a schizophrenia.

- **The principle of excluded third.** (1) That everything is either *a* or *b*: *a* + *b*; (2) that *a* and *b* are mutually exclusive (one excludes the other) and irreciprocal.

  It follows that everything is a decided and determinate thing—a simple individual that has at one time a unique determination or contradictory determinations.

- **The principle of temporal order** (*heteronomous order, irreversible order, irreflexive order*). That any two contradictory things *a* and *b* exist successively (at different times) where *a* is before *b* and *b* is after *a*: *a* < *b*. Its equivalent expression is **the principle of inequality** (broken equality, or impossible equality): (1) That everything *a* is unequal to everything else *b*; (2) that if any two things *a* and *b* are different, then they are unequal: ($a \neq b$) → ($a < b$). From the principle of temporal order, we deduce **the principle of comparability**—that everything *a* is either smaller or greater than something else *b*—and **the principle of determinate causality** (*heteronomous causality, external causality*)—that everything *a* either implies or is implied by something else *b*: *a* < *b* + *b* < *a*.

  The geometric figure expressing the analytic principles of organization one-dimensionally is the straight line. Indeed, on the straight line, the opposite points *a* and *b*—say, the beginning *a* and the end *b*—are unequal (consecutive) and determinate: *a* < *b*:

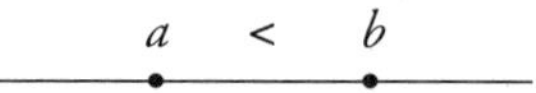

A straight line is a nonuniform, oriented, one-dimensional space with internal distinctions. Anything moving on a straight line accelerates the farther it gets away from the state of rest at the beginning *a* and approaches the state of maximum motion at the end *b*. Rectilinear acceleration is the product of a unique force generating imbalance.

If we negate the above four analytic principles, we obtain the following four synthetic principles belonging to the ideal faculty of synthetic, universal reason, which is the faculty of thinking of the object as a complex universe:

- **The principle of equivalence.** (1) That everything *a* is equal to something else *a′* (not–*a*) called *b*: *a* = *b*; (2) that everything real and numbered by the real 1 is the constant ratio of its different parts: $1 = a/b$; (3) that everything is both a and something else b: ab. This means that we have at one time alternative ways of being and of knowing the being; (4) that because things are both a and b, we do not need to ask why things are a rather than b or b rather than a; (5) that a and b are equivalent if, and only if, a has every property of b and b has every property of a, and thus their properties are common and universal; (6) that a and b are mutually inclusive (the one includes the other) or reciprocal; (7) that equality and communication between different things a and b belonging to different categories are necessary (Plato's principle of communicability); (8) that the equality between different things a and b is the real number 1 (Plato): $(a = b) (a \neq b) = 1$; (9) that unity and communication between different things is love and joy (with respect to feeling), intelligence and truth (with respect to comprehension), justice and wisdom (with respect to action).

- **The principle of reversible temporal order.** That to every temporal order there corresponds a counter-temporal order: $(a < b)(b < a)$. From the principle of reversible temporal order we infer: (1) **The principle of uncomparability**— that everything a is both smaller and greater than something else b, which is equal to asserting that everything thing a is neither smaller nor greater than something else b: $(a < b)'(b < a)'$; and (2) **the principle of indeterminate causality**—that everything a and b implies and is implied by everything: $(ab < ba)(ab < ba)$.

- **The principle of reflexive order** (*self-order or spontaneous order*). (1) That everything a is unequal to itself: $(a < a)$; (2) that everything a is both smaller than and greater than itself, a contained part and a containing whole (*self-containment*), before and after itself: $(a < a)(a < a)$. From the principle of reflexive order we derive **the principle of self-causality** (*causa sui—immanent causality*): that everything a is cause and effect of itself.

- **The principle of included third.** (1) That everything is neither (not-either) a nor b: $(a + b)' = a'b'$; (2) that everything is an impartial, undecided and indeterminate whole—a complex universe that receives at one time contrary determinations.

The geometric figure expressing the synthetic principles of organization one-dimensionally is the circular line. For example, on the circle, the opposite points *a* and *b*, the beginning *a* and the end *b*, are equal (simultaneous) and indeterminate: $a = b$:

$$a = b$$

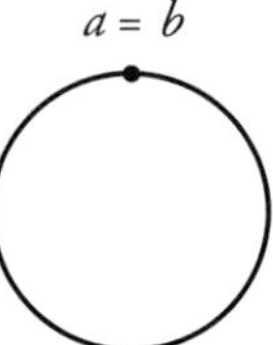

A circular line is a uniform, non-oriented, one-dimensional space with no internal distinctions. Anything moving on a circular line is both always and never at the beginning *a* and at the end *b*. Circular motion is the synthetic product or constant ratio of contrary (equal and opposite) forces generating balance.

The analytic principles of organization govern the sensible universe, which is the observable part of the physical universe revealed by the observable part of physical light. From the one-dimensional point of view, the sensible universe is an unlimited (Euclidean or hyperbolic) line without a limiting point to limit the unlimited line and to unite its consecutive, isolated parts—the individuals. An unlimited line deprived of a limiting point is an indefinitely extending (infinitely regressing) line.

The sensible universe shown by the observable part of physical light is the kind of universe that our brain perceives when it is at the state of rest. The synthetic principles of organization, on the other hand, govern the physical universe itself (τό καθαυτό) as revealed by the entire physical light. The physical universe is a complex, infinite whole—an unlimited line with a limiting point to limit the unlimited line and unite its parts (the individuals) into simultaneous, communicating wholes (the universes). The physical universe is the kind of universe that our brain perceives when it is in the state of continuous rotation at the maximum speed of light.

Quantity. We distinguish two kind of quantities:

• The sensible quantity q, which is the sensible (observable) part of the real, physical, quantity of a given kind, has at one moment a unique, finite magnitude, which we take as a unit—as equal to the finite 1: $q = 1$.

- The ideal, real, physical quantity q of a given kind, which we assimilate with the physical universe, has at a one moment a maximum, finite magnitude equal to 1 and defined as the product or constant ratio of infinite and zero magnitudes: $q = 1 = \infty \times 0$, $q = 1 = 0/\infty$; we can also call it *first quantity*.

  The place of the sensible quantity $q = 1$ is the physical universe's center a (us), whereas the place of the first quantity $q = 1 = \infty \times 0$ is the limiting-circumference b of the physical universe occurring at a maximum distance from us. This means that any finite, sensible quantity becomes at the limiting circumference of the universe, an ideal, real, physical quantity—a first quantity. The number of the first quantity is the real 1.

Relative. (1) The thing relative to something else or as something else. (2) The variable. (3) The dependent (or conditioned). See also **accident.**

Relativistic relation. The relation that measures the effect of the body's motion on its quantity: $q' = q \times \sqrt{1 - v^2/c^2}$, and $q' = q / \sqrt{1 - v^2/c^2}$, where q is the quantity of a given kind at rest and q' is the same quantity in motion. The effect of motion on quantity is studied by the theory of relativity. Its extension includes the effect of motion on perception, particularly on the energy content of the moving brain, and hence on its faculty of sensibility.

When the brain is at rest or at indefinite acceleration and its electrical energy content is lowest, it perceives only a finite part of physical light, called sensible light, which reveals only the inner, past part of the physical whole—that which we call the sensible universe. This sensible universe is conditioned by time and ruled by analytic principles of organization. The low electrical energy in the brain is manifested as its mental faculties of analytic thought and particular sensibility. When the brain moves at the maximum speed of light (or when it maximally concentrates at the limiting-point b of zero radius) and its energy content is highest, it perceives the entire physical light, which reveals the entire physical universe, its inner and outer parts, its past and future. This physical universe is timeless and ruled by synthetic principles of organization. The high electrical energy in the brain is manifested as its mental faculties of synthetic thought and universal sensibility. See also **mind.**

Real 1. The limit and total sum 1 of an unlimited series of parts. The real 1 is both a number and a principle. Taken as a number, the real 1, defined as the product or constant ratio of infinite and zero magnitudes, counts the real, physical body and measures any of its fundamental quantities m, l, t and any of their derived combinations. In this sense, the real 1 is a universal, constant, and pure number being unconditioned by our particular senses, by our states of being and motion affecting our way of perceiving things. Taken as a principle, the real 1 is the synthetic equivalence principle uniting different and distant parts of the physical body—for example, its unlimited series of parts a_n with the limit 1: $a_n = 1$.

Realism. The ontological doctrine asserting that (1) everything—namely, the universe—exists and continues to exist at all times; (2) that everything in the universe is a universe that exists and continues to exist at all times. The number of the universe is the real, infinite whole 1 defined as the product or constant ratio of infinite and zero magnitudes: $1 = \infty \times 0$, or $1 = 0/\infty$. The origin of realism is the Greek, Ionian philosophy (sixth century BCE). See its opposite **nihilism** (ontological).

Skepticism. See **nihilism**.

Scientific metaphysics. See **metaphysics**.

Sensible light. See **light**.

Sensible universe. See **universe**.

Singularity. See **cosmic singularity and organic singularity**.

Spherical geometry. The geometry of the real, physical universe, which is a complex, infinite whole composed of complex, infinite wholes having at any instant contrary determinations or the totality of determinations, and ruled by synthetic principles of being. In the spherical universe, the center a and the limiting-circumference b are both everywhere and nowhere because they are equal. The radius of the spherical universe is unlimited with a limit (complex infinite, actual infinite); when seen from inside (the center a), the spherical universe has an infinite radius containing an infinity of dimensions, and when seen from outside (the outer limiting-circumference b) at dimension zero, it has a zero radius that delimits and circumscribes its inner infinity. Thus, the spherical universe is an open-closed body having a maximum radius r equal to the real $1 = \infty \times 0$. The spherical universe composed of equal points is completely uniform in its space and time dimensions.

Spherical world (space-time). (1) The power to inverse and at the same time not to inverse the body's sense a as it moves away from the starting point. (2) The supposition that as we progress in the world, we have at any time the freedom to recede from the starting point in a straight (or hyperbolic) line or to return to the starting point in a circular line.

Substance, essence (ουσία, καθαυτό). That which *is* (τό τί ἐστιν) at all times independent of particular experience and, unconditioned by the principle of temporal order, regarded as the principle of corruption. The fundamental property of substance is therefore independence and permanence. The substance of the being is according to Aristotle, the individual obeying analytic principles of organization; according to Plato, the universe obeying synthetic principles of organization.

Sum. An arithmetic multiplicity without contact between its parts: $S = a + b$, where S is the sum and a and b are any two mutually exclusive (contradictory), isolated parts (individuals). The arithmetic sum is incomplete because it lacks the unity, contact, and continuous, immediate communication between its parts.

Supreme being (*ens summum, ideal* or *divine being*). The highest being and the highest unity of all beings. We equate the supreme being's greatest body with the physical universe, which is an infinite whole governed by the synthetic equivalence principle. The geometric topos of the supreme being is the limiting-circumference b of the physical universe occurring at a maximum distance from us. As the highest unity of all beings, the supreme being is the highest intelligence and truth, the highest justice and good, the highest love and beauty. See also **first being.**

The supreme being is both a finite individual belonging to the Euclidean series of finite individuals and an infinite whole or universe lying beyond the Euclidean series of finite individuals. This constitutes the scientific or mathematical definition of the supreme being.

Synthetic logic. The science of the principles that govern the object, thought of as a universe admitting at one moment contrary (equal and opposite) senses, which mutually neutralized generate freedom from time and sense regarded as the arrow of time.

Synthetic ontology. See **ontology.**

Synthetic reason. See **mind** (faculties of).

Synthetic universal proposition. An indeterminate, non-categorical proposition of the form: every x is both a and a' and neither a nor a'. This synthetic proposition both affirms and denies, and at the same time neither affirms nor denies, something of the object universally without absurdity; the object is thought of as a real, infinite whole 1, as a complex universe, having at one time contrary determinations (senses) and verifying the synthetic principles of equivalence and included third.

Transcendent (Kantian term). (1) What lies beyond the Euclidean, sensible world of actual and possible experience. (2) The simple limiting-point b (cosmic singularity), which is uniquely intelligible (supersensible) and intellectual (supersensuous, noumenal).

Transcendental (Kantian term). (1) What lies beyond the Euclidean, sensible world of actual experience but can be realized in the Euclidean, sensible world of possible experience. (2) The complex, limiting-point b (cosmic singularity), which is both intelligible and sensible, intellectual and sensuous. See also **cosmic singularity**.

Transcendental imagination. See **mind** (faculties of).

Uniformity. An essential property of the spherical body of the physical universe governed by the synthetic equivalence principle. A perfectly uniform universe is *homogeneous,* being the same everywhere at all points; *isotropic,* being the same in all directions; and *eternal,* being the same at all times. In the completely uniform, spherical universe, space and time are equivalent. If the universe is incompletely uniform (for example, uniform with respect to space but nonuniform with respect to time and therefore varying with time), then it is an improper universe. The incompletely uniform universe is, in fact, a Euclidean, time-conditioned, observable part of the completely uniform and timeless, spherical universe.

Universal. That which is common to different things and hence is both the same and different, one and many—the unity of multiplicity. The principles of science and nature are universal because the physical universe is completely uniform with respect to space and time. Universal principles are the same and constant everywhere, at all points, in all directions, and at all times. Universal principles that vary with time in an incompletely uniform universe are improper or incompletely universal principles of nature.

Universal sensibility. See **mind.**

Universe. According to our different ways of knowing the universe relative to our different states of being and different kinds of thought and sensibility, we divide the universe into physical universe and sensible (observable) universe.

- **Physical universe.** (1) The sum total of all things governed and numbered by the real 1 defined as the product or constant and balanced ratio of infinite and zero magnitudes:$1 = \infty \times 0$, $1 = 0/\infty$. (2) The maximum reality (*ens realissimum*), defined as the limit and total sum of an unlimited series of partial realities. (3) The real, infinite whole 1 existing in itself (τό καθαυτό) independently of our particular experience. (4) Anything that is both one thing and its inverse—the infinitely many things and nothing (the trinitarian nature of the universe). (5) The set containing everything there is and there is not. (6) Anything that admits spatial contrariety—namely, contrary determinations (senses, parts) a and $b = d'$ at the same time without absurdity, paradox, or schizophrenia. It is the kind of universe that we *ought* to perceive, when our brain is elevated to the cosmic singularity level of highest unity, energy, and speed at the physical universe's limiting-circumference b.

- **Sensible universe** (observable universe). The partial sum of things a_n indefinitely converging toward 1, which is the real, physical universe. Because the sensible universe is the *effect* of our particular perception, it is identified with the *observable* part of the physical universe. Indeed, our resting, individual brain selects (through its low-energy individual cells) only a finite portion of light, called *sensible* (observable) *light*, revealing only a finite inner and past part of the physical universe, called *sensible universe*.

Whole (or universe). The sum total, or the sum and product, of all its parts: $1 = (a + b)(ab)$, where 1 is the whole and a and b are any two distant and different parts in immediate contact. The whole is therefore a geometric multiplicity with immediate contact and unity between its parts. The whole is complete because it lacks nothing and contains parts that have continuous, immediate communication. The (proper) whole is observer-independent and unconditioned by time. The metaphysical equivalents of the whole are body, being, thing, substance. The geometric figure of the whole is the complex, indeterminate sphere of center *a* and unit radius *ab* equal to the real $1 = \infty \times 0$.

Yoga (yoke). Unification of the individual self with the supreme being, thought of as the highest being and the highest unity of all beings. The unification is possible through mental concentration on a limiting point of zero magnitude that realizes the unity of all points of the Euclidean plane. This limiting point (cosmic singularity) occurs in zero dimension both beyond and within the Euclidean plane. Attaining the maximum unity of the world is equivalent to attaining the state of supreme being—that is, the state of being a physical universe: a universal being. See also **supreme being.**

Acknowledgments

I express my deep gratitude to Georges Comtesse for his unwavering love and persistent encouragement during the long and painful times of conceiving and constructing this work.

Index